Avail of Aerosolised Ketaject vs Brine in Thwarting Recuperative Phase Excoriated Gullet

Nike Diosa

Avail of Aerosolised Ketaject vs Brine in Thwarting Recuperative Phase Excoriated Gullet

First Edition August 2025

Written by Nike Diosa

CONTENTS

S.NO	TOPIC	PAGE NUMBER
1	INTRODUCTION	1
2	AIM AND OBJECTIVES	4
3	REVIEW OF LITERATURE	5
4	MATERIALS AND METHODOLOGY	29
5	OBSERVATION AND RESULTS	38
6	DISCUSSION	54
7	LIMITATIONS	59
8	CONCLUSION	60
9	SUMMARY	61
10	REFERENCES	66

INTRODUCTION

Postoperative sore throat (POST) is the most prevalent minor complication of general anaesthesia (GA) with endotracheal intubation[1,2]. Even though it is considered as a minor complication, it can cause significant morbidity and discomfort postoperatively[3].

POST ranked as the eighth most adverse effect following general anaesthesia, with an incidence of 21 to 65%[3,4]. Many factors have been found to be responsible for post operative sore throat, and it depends primarily on the method of airway management[5]. It is frequently reported after tracheal intubation (45.4%), whereas in supraglottic airway the incidence is found to be lower (17.5%) and it is even less (3.3%), when a facemask is used as a mode of anaesthesia. The incidence and severity of post operative sore throat is found to be more in female sex[1].

Various techniques, both pharmacological and non-pharmacological, have been found to be effective in treating post-operative sore throat (POST). Appropriately sized endotracheal tubes (ETTs)[6] smooth laryngoscopy, maintaining an appropriate cuff pressure [7] lubricating the tubes with jelly[8] gentle oral suctioning, and extubation after full deflation of the cuff have been found to be effective in reducing the incidence and severity of sore throat.

Despite the fact that numerous preventive methods have been identified, none has garnered significant interest. However, one of the most effective method is considered to be drug-based preventive treatment.

Diverse pharmacological agents, including steroids dexamethasone, methyl prednisolone, fluticasone, and betamethasone [9–12] local anaesthetic agents opioids[13]nonsteroidal anti-inflammatory drugs (NSAIDs)[14]alpha agonists[15] and ketamine[16] were used to reduce the intensity and severity of postoperative sore throat after general anaesthesia.

Aspirin (NSAID) gargles has been useful in reducing post operative sore throat (POST)[14] Preoperative magnesium (N-methyl-D-aspartate receptor antagonist [NMDA]) lozenges have been discovered to reduce the incidence and severity of post operative sore throat (POST)[17]

Magnesium sulphate topical preparations are well-known for their therapeutic and preventative effects on thrombophlebitic tissue[11]

Ketamine is an N-methyl d-aspartate (NMDA) receptor antagonist used in multiple preparations ranging from gargles to nebulisation. It has been shown to be effective in preventing postoperative sore throat[18].It have anti-nociceptive and anti-inflammatory effects[19].

Ketamine nebulization has the following advantages over other modes of administration: It reduces the bitter taste of ketamine, a smaller volume can be useful in nebulisation than in gargle, where a larger volume with a diluted concentration has been found to be beneficial, and there will be a reduced risk of aspiration due to the lower volume. It also benefits from improved patient cooperation.

In our study we evaluated the efficacy of nebulisation with ketamine and normal saline given 15 min prior to general anaesthesia helps in prevention of post operative sore throat.

REVIEW OF THE LITERATURE

Edward A. Loeser et al[2] (1980) examined the effect of tracheal tube lubrication versus cuff design on the incidence and severity of postoperative sore throat in six groups of twenty patients. Group 1 to 5 had tracheal intubation with Portex disposable polyvinylchloride tracheal tubes and group 6 had mask anaesthesia. This study result shows that lubrication of tracheal tubes provides no advantage of reducing postoperative sore throat and it can even increase the incidence and severity of sore throat in some cases, hence concluded that the tracheal tube causing the least postoperative sore throat is unlubricated low residual volume cuff.

O. Canbay et al[18] (2008) conducted a prospective, randomised, placebo-controlled, single-blind study on the effectiveness of ketamine gargle in reducing postoperative sore throat. Two groups of 23 patients were randomly assigned: Group C received 30 ml of saline; in the ketamine group 30 ml of saline was mixed with 40 mg of ketamine and patients were instructed to gargle for 30 seconds, 5 minutes prior to induction of general anaesthesia. POST occurred more frequently in the saline group compared to the ketamine group at various postoperative time intervals, and significantly more patients in the saline group suffered from severe sore throat compared to the ketamine group ($P<0.05$); therefore, ketamine gargle significantly decrease the incidence and severity of POST.

Prerana P Shroff et al[23] (2009) conducted a prospective randomized, controlled study over 3 months in 150 patients which was divided into three equal groups (air, isotonic saline and alkalinized lignocaine as inflation media) undergoing general anaesthesia with controlled ventilation with nitrous oxide and oxygen. Result of this study shows a volume of air increased just before extubation in the air group and also shows that the tube intolerance, hoarseness and postoperative sore throat were lesser with lignocaine group

Zhuquan Su et al[26] (2017) aimed to establish an animal model and to investigate risk factors associated with post intubation tracheal stenosis. They intubated with 4 sized tubes with an internal diameter of 6.5 to 8.0 mm and cuff pressures of 100 to 200 mmHg for 24 hr. The tracheal wall was evaluated by bronchoscopic and histological examinations. The degree of tracheal stenosis ranged from 78% to 91% at 2 weeks. The histopathological pictures are similar to those of clinical cases of postintubation tracheal stenosis and hence this study concluded by saying excessive cuff pressure and bigger sized tube are associated risk factors for postintubation tracheal stenosis.

Raham et al[21] (2018) used nebulisation as a mode of medication delivery as a pre-emptive strategy to alleviate a Postoperative Sore Throat (POST) in children receiving general anaesthesia. It is a prospective, randomized, double-blind study involving 36 patients of 6-16 years divided into 3 groups of 36 patients each. Preoperative nebulization of 40 mg/kg magnesium sulphate, 1 mg/kg of ketamine,0.16 mg of dexamethasone and he concluded that the incidence of post-operative sore throat is significantly reduced with a p value of <0.003 at 4th hour in ketamine group than with any other groups while the dexamethasone group shows a additional advantage of no incidence of vomiting and decreased incidence of post-operative cough.

M. Reddy et al[22] (2018) conducted a Dose-dependent study on effectiveness of ketamine nebulisation in preventing post-operative sore throat following tracheal intubation. The primary objective was to assess graded doses of ketamine with 0.5mg/kg, 1mg/kg and 1.5mg/kg given on various time interval 5 minutes and 15 minutes. In the result it is observed that 1mg/kg and 1.5mg/kg of ketamine showed statistically significant same effect on postoperative sore throat, however larger population studies are needed with the estimation of serum ketamine levels to find out a better dose of ketamine for nebulization to prevent post-operative sore throat (POST).

Yasser Hammad et al[24] (2019) examine the effects of Endotracheal Tube Size and Cuff Pressure on the Incidence of Postoperative Sore Throat at three distinct tertiary care institutions using a sample size of 100. Endotracheal tube cuff pressure and sizes were measured by the observer after induction of general anaesthesia. The sore throat was observed after full recovery of the patients and before discharge from PACU by another observer. Result shows that the sore throat was not significantly correlated to either tube size or cuff pressure.

Efram fenta et al[25] (2020) conducted a study on Incidence and factors associated with postoperative sore throat for patients undergoing surgery under general anaesthesia with endotracheal intubation on 123 participants, out of which, 60 were developed Post-operative sore throat. The multivariate logistic regression of this study showed that sex, size of the endotracheal tube and number of attempt had statistically significant association with POST and it also revealed that larger size of ETT were associated with the causative factor of the Postoperative sore throat with p values of 0.01, 0.028, and 0.031 for ETT sizes of 6.5, 7.0 and 7.5 respectively and hence this study concluded that the smaller ETT size and reduced number of attempts of laryngoscopy minimises the incidence of POST.

Neepa et al[20] (2022) conducted a study titled "Comparative evaluation of incidence of post-operative sore throat after nebulization with ketamine and magnesium sulphate in patients undergoing general anaesthesia requiring endotracheal intubation in patients of ASA grading I- III, aged between 18-65 years on 88 patients of ASA grading I- II, out of which 44 received preoperative nebulization with 50mg of ketamine with 1ml of normal saline while others group received 500mg Magnesium sulphate with 1ml of normal saline and he observed the post-operative sore throat, hoarseness of voice and cough in various time intervals in which the result shows significant reduction of sore throat in group ketamine when compared to magnesium sulphate group at 0hr (p=<0.01), 2hr (p=<0.01), and 4hr (p=<0.01).

ENDOTRACHEAL TUBE:[27–33]

The Endotracheal tube is a device that is inserted through the larynx into the trachea to deliver the gases to and from the lungs.

General Principles:

Resistance and Work of Breathing:

The tracheal tube caused more resistance in case of spontaneously breathing patient and hence it is important factor in determining work of breathing than the breathing system itself. This concern is more in case of paediatric patients. Several factors are there to determine about the resistance offered by the endotracheal tube.

Internal Diameter:

The most important factor in determining the resistance is internal diameter (ID) of the endotracheal tube. The resistance will be more in thick-walled tubes than thinned wall tubes with the same outer diameter (OD). When the tube size decreases wall thickness to tube diameter ratio is greater thus offering more resistance to the air flow. Resistance will be increased in case of insertion of further device such as suction catheter, fiberscope.

Length:

Decreasing the length of the tube offers lesser resistance than the longer tubes.

Configuration:

Curved tube or connector, in case of any kinking of the endotracheal tube offers more resistance. The resistance of swivel and curved connectors are almost same.

Dead Space:

Dead space is the part of the respiratory system which doesn't take part in the gaseous exchange. The tracheal tube offers the mechanical dead space but the dead space which is offered by the volume of a tracheal tube is lesser than the dead space, which is offered by the natural passages, but in case of paediatric patient longer tube and the connectors will increase the dead space.

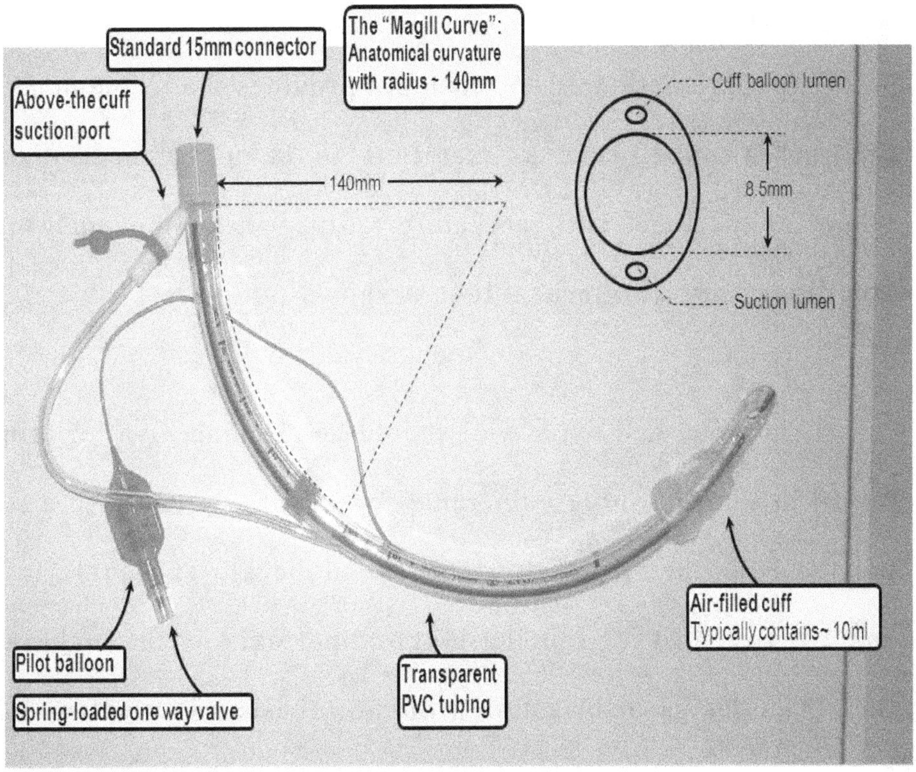

FIGURE 1: PARTS OF ENDOTRACHEAL TUBE

Tracheal Tube System:

Polyvinyl chloride tubes is the most widely used disposable tracheal tubes. It is comparatively inexpensive and better compatible with the tissues. These tubes are less chances of kinking than the conventional red rubber tube. They are stiff enough and softens on body temperature, hence conforms to the respiratory passages. PVC tubes also facilitates the passage of a suction catheter or bronchoscope. Their transparency of the tube also permits the fogging of the tube confirming the placement of the endotracheal tube and helps in case of any secretion through the endotracheal tube.

Tube Design :

The international/U.S. standard provides the recommendations on tracheal tubes such as materials to be used, inner diameter, length of the tube, cuff, curvature radius, packaging, marking and labelling. A typical tracheal tube is shown in Figure 1

The inner and outer wall should be circular. Oval or elliptical lumen in shape is more vulnerable for kinking. They have a slanted portion near the tip of the tube called bevel. The angle is acute angle between bevel and the longitudinal axis of the endotracheal tube. The design or orientation of the tube is important when it is introduced over a fibreoptic scope. For nasotracheal intubation hemispherical bevel reduces the nasal injury. During nasal intubation, bevel facing upwards helps in the prevention of injury to the epiglottis during the advancement. Murphy's eye is the hole, which is present opposite to the bevel, it provides the alternate pathway for the gas flow when the bevel is occluded. This provides the additional advantage if the tube enters the mainstem bronchus. Tracheal tubes which lack the murphy's eye is known as Magill tubes. The American Society for Testing and Materials/International Standards Organization standard requires that a radiopaque marker is placed at the patient end or the entire tube for finding the tube position.

Cuff Systems:

A cuff system consists of the cuff with a inflation system with inflation tube, a pilot balloon, and an inflation valve. An inflation lumen in the tube wall may also be present. The purpose of the cuff system is to provide a seal between the tube and tracheal wall to prevent passage of pharyngeal contents into the trachea and ensure that no gas leaks past the cuff during positive-pressure ventilation. The cuff also serves to centre the tube in the trachea.

Cuff:

The cuff is an inflatable pouch present in the distal end of the tube. The cuff material should be soft and tear resistance at the same time it should be soft.

Cuff Types: It is classified based on the volume and the pressure of the cuff.

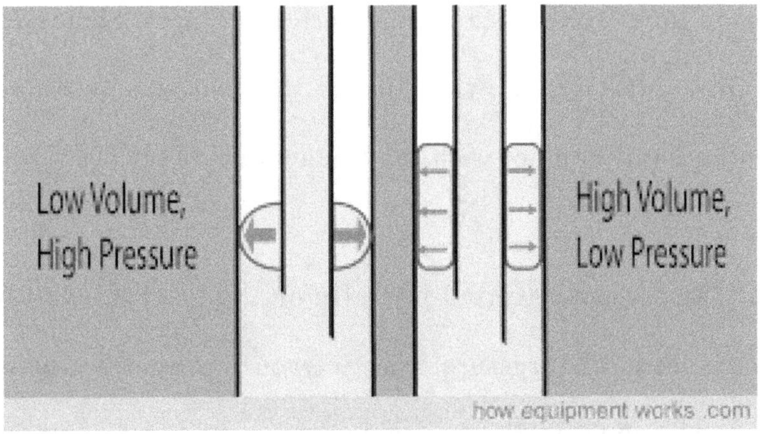

FIGURE 2: PICTORIAL REPRESENTATION OF TWO TYPES OF CUFFS.

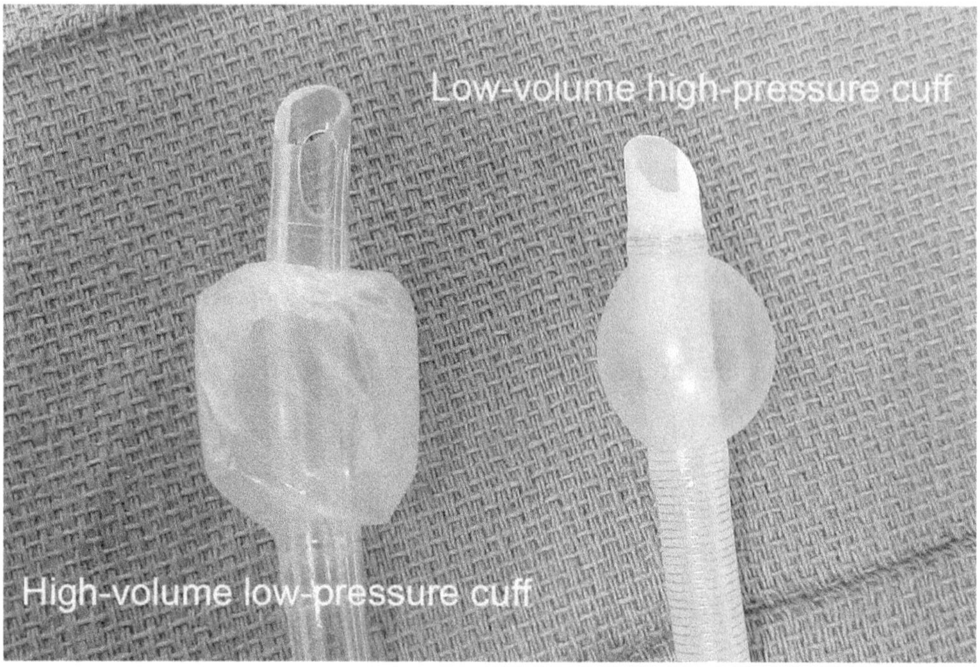

FIGURE 3: HIGH VOLUME LOW PRESSURE AND LOW VOLUME HIGH PRESSURE CUFF AFTER INFLATION

Low-volume, High-pressure Cuff:

The low-volume high-pressure cuff has a small diameter with low residual volume (the amount of air that can be withdrawn from the cuff after it has been allowed to assume its shape with the inflation tube exposed to atmospheric pressure). Hence it requires a high intra cuff pressure to maintain the seal and its property. It has a small area of contact with the tracheal mucosa.

The pressure exerted laterally on the tracheal wall will be less than the intracuff pressure. The intracuff pressure does not change when the tracheal wall is contacted.

Advantages:

1. These cuffs provide better prevention from aspiration and good visibility during intubation.
2. Incidence of sore throat is also lower with this type of cuff.
3. They are reusable and not expensive, hence can be used as a routine in case of adult.

Disadvantages:

1. The most common disadvantages is ischemic injury to the tracheal wall due to its prolonged use.
2. If the surgical time exceeds for several hours, tube should be replaced with the high volume and low-pressure cuff.

High-volume, Low-pressure Cuff:

A high-volume, low-pressure cuff has a increased resting volume and compliance, because of this property it can provide a better seal without much increase in pressure. As the cuff get inflated the widest part will first touches the trachea, as it continuously get inflated the cuff area which is coming into contact also increased. If inflation of the cuff continues, the area of the trachea which is in contact with the cuff will have a tracheal distortion.

A cuff with a short tracheal contact area will have a lower incidence of postoperative sore throat. The intracuff pressure varies during the ventilatory cycle. In a spontaneous breathing patient, cuff pressure will be negative during inspiration and positive during expiration. In controlled ventilation, when airway pressure increased than intracuff pressure, positive pressure falls on lower part of the cuff, since the cuff wall is pliable, it will be unable to resist the pressure and it get changes into a cone shape as proximal part distended and distal part is compressed. The air in the cuff will get compressed to the point where intracuff pressure equals with a airway pressure. During exhalation, the intracuff pressure will decrease until its resting pressure.

Advantages:

1. If the cuff is not stretched the intracuff pressure will be almost approximate with the pressure on the tracheal wall hence it is easy to regulate the pressure exerted on the tracheal mucosa.
2. The most common cuff induced complication like tracheal stenosis and postoperative sore throat can be reduced.

Disadvantages:

1. This type of tube may obscure the view of the glottis during the insertion.
2. When forceps are used, it might easily get torn.
3. It has a higher chance of dislodgement also even while applying the negative pressure in the nasogastric tube.

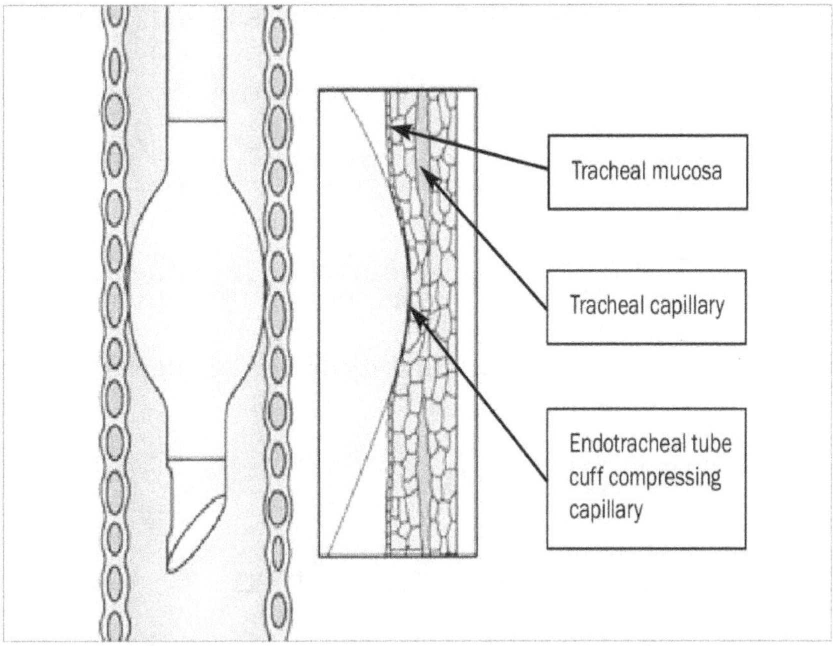

FIGURE 4: EFFECT OF CUFF ON TRACHEAL MUCOSA.

Even when the cuff pressure is high as 60mmhg it will not provide a safety against the fluid from leaking into trachea. It is more with spontaneous ventilation than the positive pressure ventilation, continuous positive pressure ventilation, pressure-controlled ventilation.

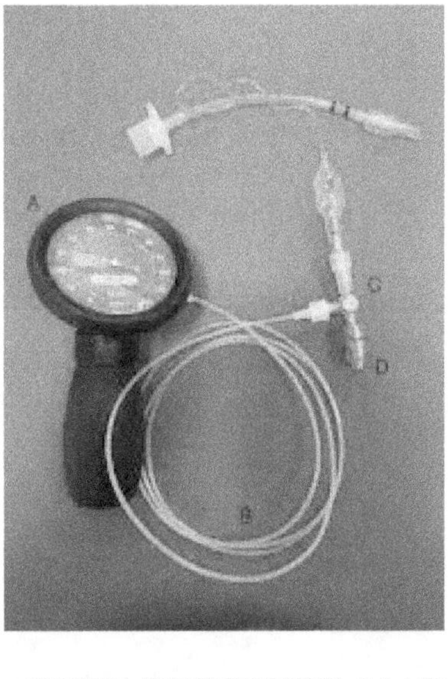

FIGURE 5: CUFF PRESSURE MANOMETER.

Lubrication the tube with the water-soluble gel and silicone cuff with constant pressure inflation system will prevent leakage for certain period of time.

During maintenance of anaesthesia, the nitrous oxide that we use will diffuse into the tube and it further adds up to the volume of the cuff.

Sore throat following tracheal intubation:

The highest incidence of post-operative sore throat occurs with the patient who have undergone endotracheal intubation. In 1325 patients, the incidence of sore throat was 14.4%. In this study, for women 8mm and men 9mm tracheal tube along with the lubrication using lignocaine jelly. However, the incidence of sore throat in women (17%) was significantly higher than that in men (9%). Multiple attempts at intubation did not increase the incidence of sore throat.

It has been evident that use of the smaller size tracheal tube will have a decreased incidence of post-operative sore throat due to the low area of contact with the tracheal surface.

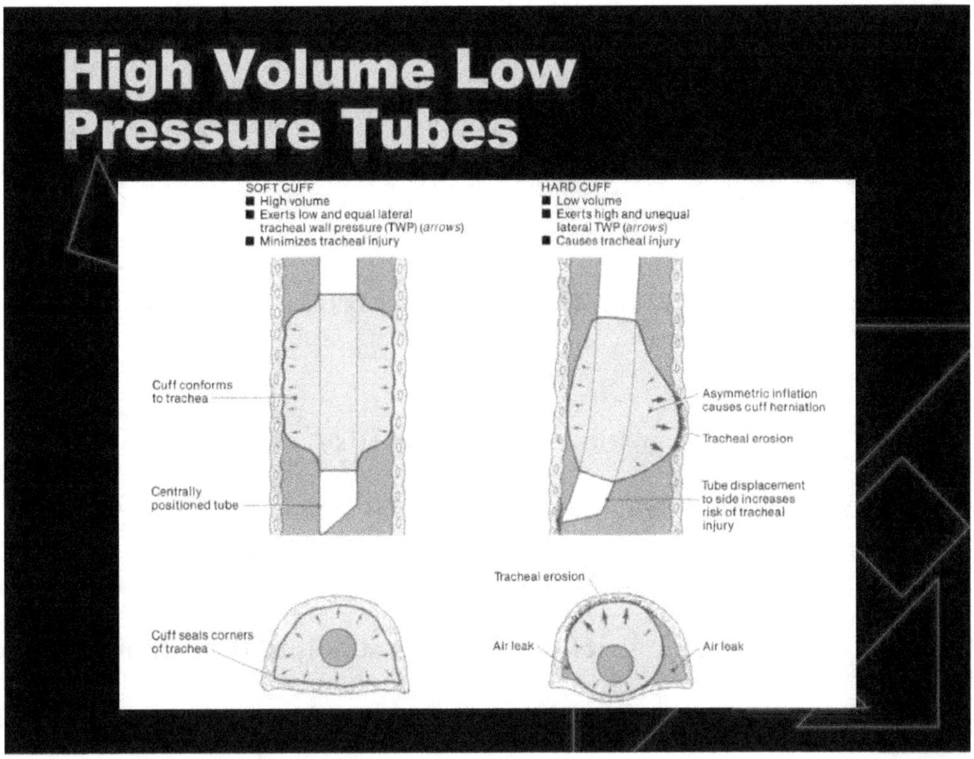

FIGURE 6: EFFECT OF HARD CUFF AND SOFT CUFF ON TRACHEAL MUCOSA.

The cuff pressure is the major factor which influences post-operative sore throat. The red rubber tube which has a low residual volume and high pressure which it will exert on the tracheal mucosa and can cause severe sore throat following extubation.

A study of blood flow in rabbit tracheal mucosa concludes that the high pressure low volume cuff when used for a long period of time it may cause tracheal mucosa ischemia and it further leads to tracheal stenosis and tracheomalacia but when the thin walled, low pressure high volume cuff is used it doesn't impair the blood flow even pressure upto 80-120mmhg, it is mainly because of equal exertion of pressure in trachea in various areas. Because of all these factors, recommendation was maintaining a cuff pressure of < 20 mmHg. Recommended ideal cuff should have a diameter slightly less than trachea and material that would allow a 10% increase in diameter over the range of inflating pressure of 20–30 cm of water.

Lubrication with 1% hydrocortisone was also found to increase the incidence of sore throat from 50% to 90%, when compared with KY jelly. The use of lubricating jelly containing a local anaesthetic is beneficial in the reduction of postoperative sore throat after tracheal intubation. The role of succinylcholine is not clear but it is known for causing postoperative muscle pain and can also cause sorethroat because of the striated pharyngeal muscles.

In conclusion the use of the smaller or appropriate sized endotracheal tube with small area of contact have found to be beneficial in preventing and treating postoperative sore throat. Careful control of intracuff pressure during intraoperative periods is mandatory. Lubricants having the local anaesthetic agents are not useful and in some cases it might even increases the incidence and severity of postoperative sore throat.

Treatment of postoperative throat symptoms:

Most of the mild cases resolve spontaneously and there is no need for the additional treatment at all. In moderate to severe cases of postoperative sore throat can be treated with benzydamine hydrochloride gargle which is been found effective in the treatment of postoperative sore throat. It is a topical NSAID that also have local anaesthesia property, since it has alkaline pH it will not absorb systematically. The most common injury is hematoma of the left vocal cord which doesn't have any treatment and it resolves spontaneously. Penetrating injuries by the laryngoscopic blades and other airway instruments have to be treated with the topical antibiotics.

KETAMINE:

Ketamine is a water-soluble molecule structurally similar to the phencyclidine.

Availability: It is available in vials containing 10 mg/ml and 50 mg/ml. For central neuraxial blockade preservative free ketamine is available. It available as a racemic mixture of S and R enantiomers.

S-ketamine is 5 times more potent than the racemic mixture and has lesser pH of 4.5. The preservatives used in ketamine are benzalconium chloride (allergic tendency) or chlorobutanol (neurotoxic).

Mechanism of action: Ketamine binds non-competitively to the phencyclidine recognition site on N-methyl-d-aspartate receptors. Ketamine also have effect on opioid receptors, monoaminergic receptors, muscarinic receptors and voltage-sensitive sodium and L-type calcium channels and neuronal nicotinic acetylcholine receptors and have weak actions at GABA A receptors. Inflammatory mediators produced locally can activate neutrophils and it get released in the blood vessels. Direct inhibition of cytokines in blood by contribute to the analgesic effects.

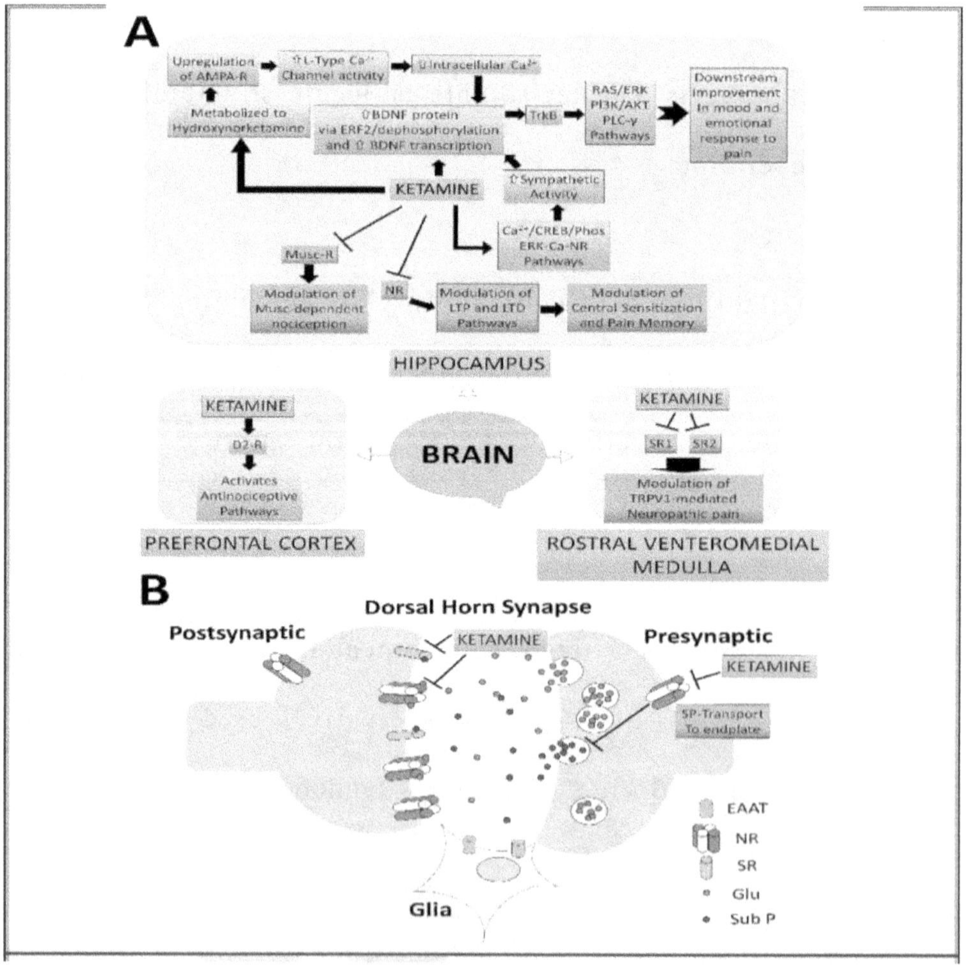

FIGURE 7: MECHANISM OF ACTION OF KETAMINE.

Antagonism/inhibition	Agonism/activation
NMDA receptors	*Opioid receptors (particularly μ, δ)*
– Dissociative anesthesia	– Central antinociception
– Amnesia	*AMPA receptor*
– Inhibited sensory perception	– Antidepressant
– Analgesia	
HCN channels	*$GABA_A$ receptors*
– Hypnosis	– Anesthetic effects
Calcium channels (L type voltage-dependent)	
– Negative cardiac inotropy	
– Airway smooth muscle relaxation	
Voltage-gated sodium channels	
– Decreased parasympathetic activity	
– Local anesthetic effect	
BK channels	
– Analgesic effect on neuropathic pain	

FIGURE 8: ACTIONS ON KETAMINE ON VARIOUS CHANNELS.

Uses, dose and route: It can used in various route such as intravenous, intramuscular, oral, intranasal and epidural.

For induction of anaesthesia: 1 - 2 mg/kg IV. It has a Broncho dilatory effect hence it is beneficial in using as a induction agent in asthmatic patients. It maintains the systemic vascular resistance and reduces the right to left shunt, hence it is used in case of tetralogy of fallot. It is also beneficial in case of hypovolemic patients.

For analgesia: The dosage is 0.5 mg/kg IV. It can used as a supplement to regional anaesthesia and it can also be used in obstetric anaesthesia.

Low dose ketamine for intra and postoperative analgesia: An initial bolus of 0.5 mg/kg of ketamine followed by a continuous infusion of 3 μg/kg/min during surgery and 1.5 μg/kg/min for 48 hours after surgery has been used successfully in total knee arthroplasty.

As a bronchodilator: It may be used to treat status asthmaticus along with other bronchodilators (30-40 μg/kg/min)

As a premedicant: 3 – 5 mg/kg IM (onset 5 min), nasal – 3-6 mg/kg (5 min); oral – 3-10 mg/kg (onset 20-45 min) or rectal 10 mg/kg. can be used as a sole anaesthetic where there is no requirement of muscle relaxant. It may be used as an infusion in the dose of 15-45 μg/kg/min with 50-70% nitrous oxide and 30-90 μg/kg/min without nitrous oxide.

Adverse effects and points to remember:

1. Ketamine are known to produce hallucinations.
2. It causes increase in secretions. Hence, administration of intravenous ketamine should always be preceded by a benzodiazepine like midazolam for the reduction of hallucinations and an antisialogogue such as glycopyrrolate.
3. For intramuscular use as a premedication, it may be mixed with glycopyrrolate to produce drying of secretions.
4. For oral premedication ketamine should be mixed with atropine and not glycopyrrolate.

NEBULISATION:[34-37]

The first pressurized inhaler was invented in France by Sales-Girons in 1858. It uses pressure to atomize the liquid. The pump handle is used like a bicycle pump. When it is pulled up, it draws liquid from the reservoir, and upon the force of the user's hand, than the liquid is pressurized through an atomizer, which is sprayed out for inhalation near the user's mouth.

FIGURE 9: PRESSURIZED NEBULISER FROM 1858

The first electrically driven gas compressor was invented in the 1930s and called a Pneumostat. With this device, a medical liquid. In 1964, an another new type of electronic nebulizer was introduced which is called as ultrasonic wave nebulizer.

A nebulizer is a drug delivery device which administer the drug in a form of mist which is then inhaled in the lungs. It is used in the treatment of bronchial asthma, cystic fibrosis, Chronic obstructive pulmonary disease and other respiratory diseases. An aerosol is a mixture of gas and solid or liquid particles.

Aerosol deposition:

The deposition of the drug in the lungs largely depends on the particle or droplet size. Smaller the particle the greater its chance of peripheral penetration and retention but for fine particles below 0.5 µm in diameter there is a chance of inadequate deposition and it will get exhaled.

In 1966 the Task Group on Lung Dynamics, concerned about the hazards of inhalation of environmental pollutions, proposed a model for deposition of particles in the lung. This states that the particles of size 10micro meter or more than that in diameter are likely to deposit in the mouth and throat whereas in case of 5–10 µm diameter a transition from mouth to airway deposition occurs, and particles smaller than 5 µm in diameter deposit more in the lower airways which are appropriate for pharmaceutical aerosols.

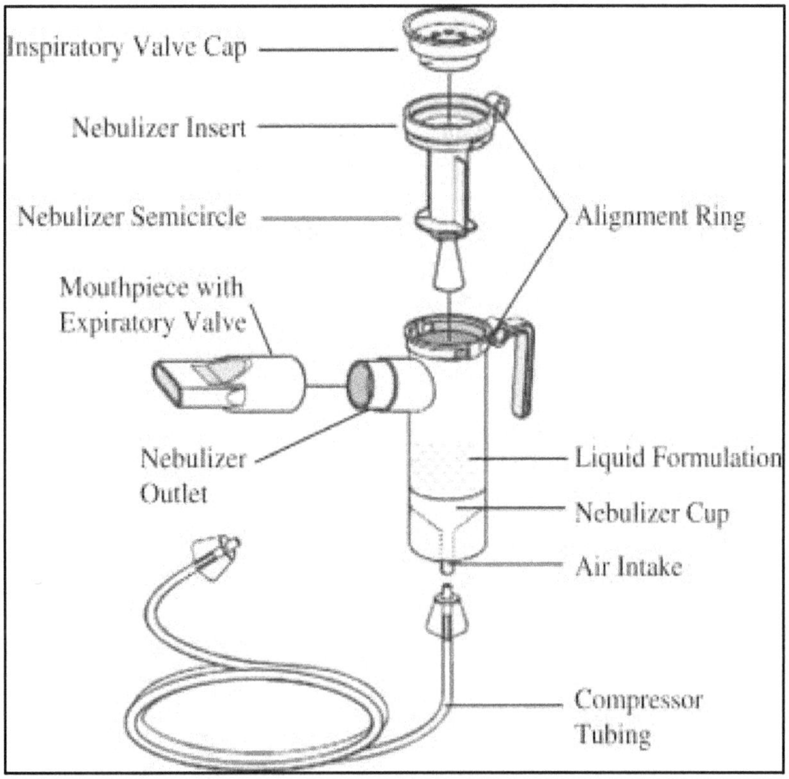

FIGURE 10: PARTS OF NEBULISER

Use and attachments:

Corticosteroids and bronchodilators such as salbutamol are used and sometimes in combination with ipratropium. Inhalational route is used due to the fact that it will target the drug in the respiratory system and reduces the side effects produced by the drug and onset is also rapid.

Normally the aerosolized medicine is inhaled through a mouthpiece, similar to that of an inhaler sometimes mouthpiece can be replaced with the face mask since it is useful in case of paediatric age group and young adults. Many nebulizers have pacifier attachments for infants. Usually, mouthpieces are preferable if patients can use since facemasks result in reduced lung delivery because of aerosol loses in through the nose.

In this study we have used face mask nebuliser as a nebulisation tool for the effective delivery of the aerosols to the patient's airway.

MATERIALS AND METHODOLOGY

Study design and setting:

This was a prospective randomized comparative study conducted in the Department of Anaesthesiology, Saveetha Medical College, Tamil Nadu.

Study duration:

The duration of the study was for 12 months after institutional ethical committee approval.

Sample size:

Sample size was estimated by using nMaster software Version 2.0 by applying details in the formula. Based on the study "Ahuja V, Mitra S, Sarna R. Nebulized ketamine decreases incidence and severity of post-operative sore throat. Indian J Anaesth 2015; 59:37-42 shows that 23 (46%) patients in group S and 10 (20%) patients in group K had POST at some point of the study.

Formula

$H_0: P_1 = P_2$, $\quad H_a: P_1 \neq P_2$

$$n = \frac{\left\{ Z_{1-\frac{\alpha}{2}} \sqrt{2\bar{P}(1-\bar{P})} + Z_{1-\beta} \sqrt{P_1(1-P_1) + P_2(1-P_2)} \right\}^2}{(P_1 - P_2)^2}$$

Where,

$$\bar{P} = \frac{P_1 + P_2}{2}$$

P_1 : Proportion in the first group

P_2 : Proportion in the second group

α : Significance level

$1-\beta$: Power

Based on the above parameter with an alpha of 0.05 (2 sided) and power of 80% the estimated sample size using the sample size formula for two proportion comparison.

Proportion in group I	0.46
Proportion in group II	0.20
Estimated risk difference	0.26
Power (1- beta) %	80
Alpha error (%)	5
1 or 2 sided	2

$$n = \frac{(1.96 * \sqrt{2 * 33 * 67} + 0.842\sqrt{46 * 54 + 20 * 80})^2}{(26)^2}$$

$$n = \frac{(1.96 * 66.49 + 0.842 * 63.90)^2}{676}$$

$$n = \frac{(130.32 + 53.80)^2}{676}$$

$$n = \frac{33900.17}{676} = 50.14 \approx \mathbf{50}$$

The above parameter and formula give us a sample size of 50 samples per group.

Methodology:

Inclusion criteria:

- ASA I and II
- Both male and female patients
- Patients age: 18 to 65 years
- Patients undergoing elective surgeries

Exclusion criteria:

- Patient refusal.
- Patients with a known allergy to the study drugs.
- Significant cardiac, respiratory and central nervous system dysfunctions.
- Anticipated airway difficulty.
- Any history of psychiatric disorders.
- Patients undergoing surgery of the head, neck, or mouth.
- Pregnant females.
- Recent anti-inflammatory drug intake.

Research design:

One day prior to surgery, all of patients underwent a preoperative evaluation consisting of a thorough medical history, physical examination, and investigations including preoperative urine analysis, blood urea, serum creatinine, complete blood count and blood sugar, liver function tests, electrocardiography (ECG) and chest X-ray.

All patients were instructed to fast for at least eight hours prior to surgery. Each patient was given an explanation of general anaesthesia (GA) and written informed consent was obtained from the patient and his family in the presence of an independent witness.

Using opaque, sealed envelopes containing computer-generated random number tables, a non-participating anaesthesiologists randomised patients into two groups. The staff nurse opened the envelopes and prepared nebulization solution according to group allocation.

Patients in Group K received ketamine (4.5ml of normal saline and 0.5ml of ketamine 25mg) while patients in Group S received 5ml of normal saline.

Patients received the study drug via nebulization mask connected to an oxygen-driven source (8 L) mounted on the wall for

15 minutes prior to surgery. The registered nurse did not take part in the subsequent evaluation of these patients. Patients were blinded and they were unaware of the nebulisation solution that were administered. After nebulization, patients were transferred to the operating room and intraoperative monitoring devices were applied, including continuous ECG, non-invasive blood pressure (BP), pulse oximetry (SpO2), and end-tidal carbon dioxide. After completing nebulization of the patients, general anaesthesia was induced 10 minutes later. The dorsum of the patient's hand was cannulated intravenously, and crystalloid infusion was initiated.

Intravenous injections of 0.2 mg glycopyrrolate and 100micrograms fentanyl were administered as pre-anaesthetic medications. 2mg/kg of propofol were administered intravenously to induce general anaesthesia. Succinylcholine 2 mg/kg was injected intravenously to facilitate tracheal intubation, and a Soft Seal cuffed sterile polyvinyl chloride tracheal tube with an internal diameter of 7–7.5 mm for women and 8–8.5 mm for men was used to intubate the trachea. Operating room anaesthesiologist performed tracheal intubation using direct laryngoscopy (Macintosh blade) after ensuring adequate neuromuscular blockade. Bougie guided intubation and patients with more than one intubation attempts were excluded from the study. The tracheal tube cuff was inflated until no air leakage could be detected with a stethoscope and cuff pressure monitoring was done. Sevoflurane was added to a mixture of 50% nitrous oxide and 50% oxygen to maintain GA. Injections of vecuronium (4 mg iv stat and 1 mg SOS) helped to maintain muscle relaxation. Injections of 4 mg ondansetron were administered 30 minutes before the end of surgery and 8 hours afterward. At the conclusion of surgery, when the patient was sufficiently sedated, the oropharynx was suctioned gently, and sevoflurane was discontinued. The oxygen concentration during inspiration was increased to 100%. After spontaneous respiration resumed, neostigmine 0.05mcgs/kg and glycopyrrolate 0.01mcgs/kg were administered intravenously to reverse the neuromuscular block. The patient's trachea was extubated when they were conscious and had full motor function. Surgeries lasting more than three hours were also excluded from the study.

Parameters observed:

1. Hemodynamic parameters including Heart rate, Blood pressure (systolic, diastolic, and mean arterial pressure), and Oxygen saturation were recorded at pre-nebulization (baseline), pre-induction, after induction, 15 minutes, 30 minutes, 45 minutes, and 60 minutes of GA.
2. Sore throat evaluation was performed immediately following extubation, 0 hours, 2 hours, 4 hours, 8 hours, 12 hours, and 24 hours after surgery.

The grade for POST was based on a four-point scale.

- 0 = no throat pain.
- 1 = mild throat pain (complaints of sore throat only on asking)
- 2 = moderate sore throat (self-reported sore throat complaints).
- 3 = severe throat pain (change of voice or hoarseness, associated with throat pain).

Other adverse effects, if any, were observed. All observations were recorded by a resident in anaesthesia who was unaware of the medications administered. For postoperative pain relief, intravenous paracetamol 1gram 8th hourly and tramadol 100 mg SOS were administered.

CONSORT 2010 Flow Diagram

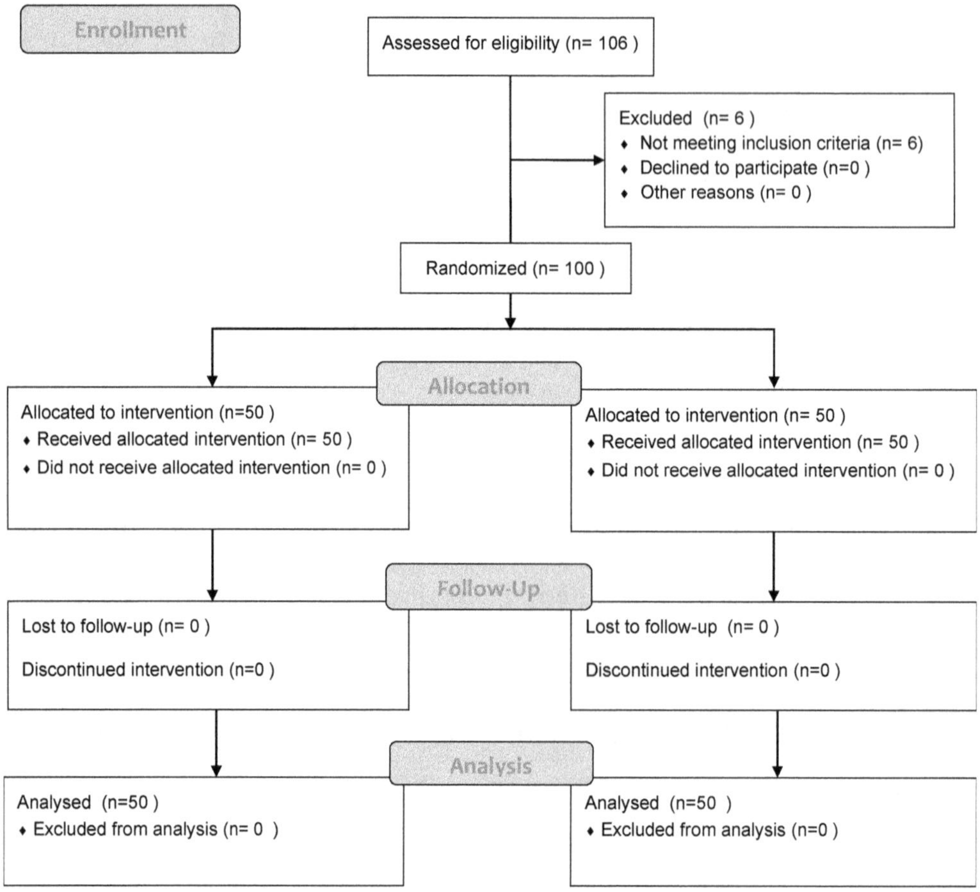

Statistical analysis:

The collected data were analysed with IBM SPSS Statistics for Windows, Version 23.0. (Armonk, NY: IBM Corp). To describe about the data descriptive statistics frequency analysis, percentage analysis was used for categorical variables and the mean & Standard deviation were used for continuous variables. To find the significant difference between the bivariate samples in independent groups the independent sample t-test and the Mann-Whitney U test was used. To find the significance in categorical data Chi-Square test was used. In all the above statistical tools the probability value 0.05 is considered as significant level.

Ethical consideration:

Informed consent was obtained from all the participants. Participant information sheet was provided with the details of the study and was explained to the participants before getting the consent. Patient's identity, privacy and confidentiality was maintained throughout the study period. Both the group had the same standard of care.

OBSERVATION AND RESULTS

Age:

There is no statistically significant differences between the group with respect to age (p value-0.289)

Table 1: Age distribution of the study participants.

			Nebulisation		Total	P- value
			Ketamine	Saline		
Age	Upto 20 yrs	Count	3	10	13	0.289
		%	6.0%	20.0%	13.0%	
	21 - 30 yrs	Count	7	6	13	
		%	14.0%	12.0%	13.0%	
	31 - 40 yrs	Count	5	8	13	
		%	10.0%	16.0%	13.0%	
	41 - 50 yrs	Count	12	8	20	
		%	24.0%	16.0%	20.0%	
	51 - 60 yrs	Count	13	8	21	
		%	26.0%	16.0%	21.0%	
	61 - 70 yrs	Count	5	7	12	
		%	10.0%	14.0%	12.0%	
	Above 70 yrs	Count	5	3	8	
		%	10.0%	6.0%	8.0%	
Total		Count	50	50	100	
		%	100.0%	100.0%	100.0%	

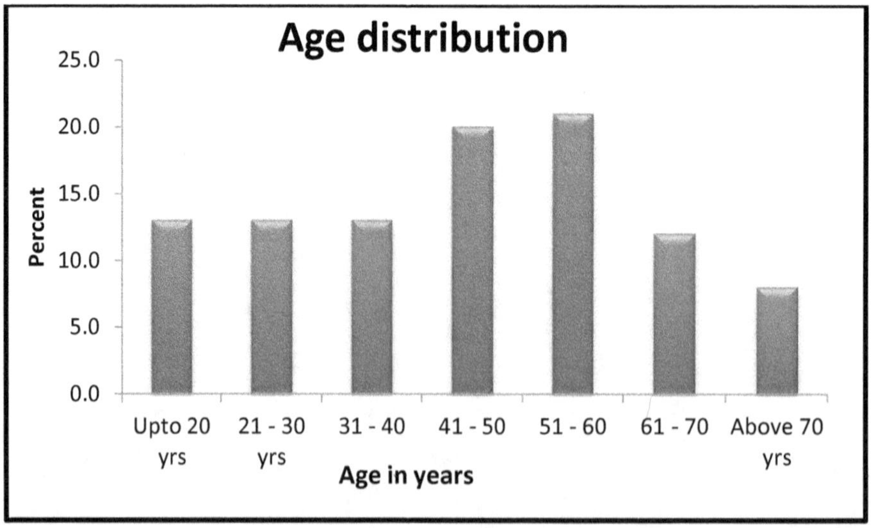

Figure 12: Age distribution of the study participants.

SEX:

Gender of the study participants were comparable between both the groups with a insignificant p value of 0.315

Table 2: Gender distribution of the study participants

			Nebulisation		Total	P value
			Ketamine	Saline		
SEX	female	Count	25	20	45	0.315
		%	50.0%	40.0%	45.0%	
	male	Count	25	30	55	
		%	50.0%	60.0%	55.0%	
Total		Count	50	50	100	
		%	100.0%	100.0%	100.0%	

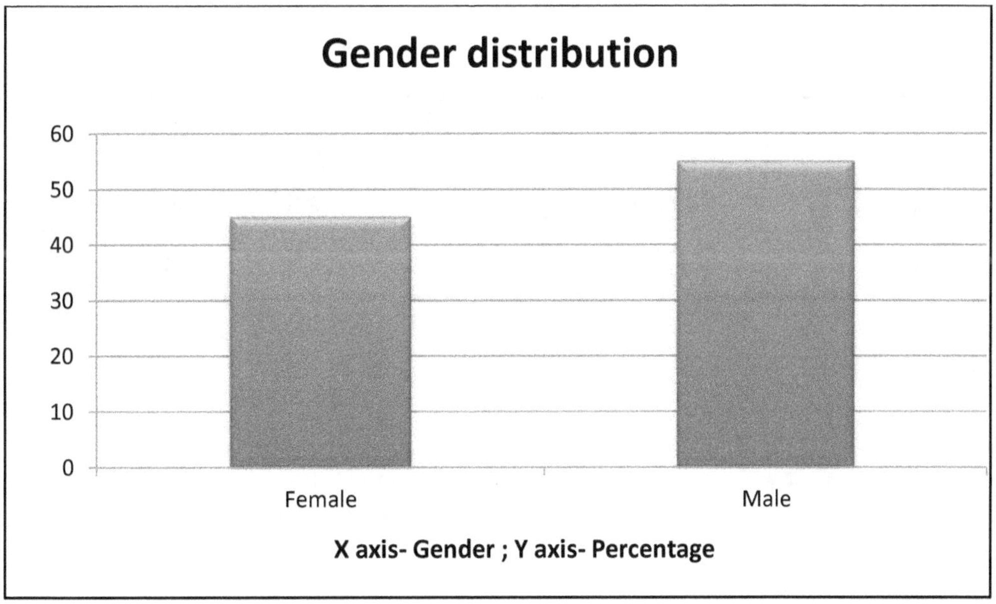

Fig 13: Gender distribution of the study participants

BODY MASS INDEX (BMI) AND DURATION OF SURGERY:

Body mass index in each group is almost similar with the mean BMI of ketamine group is 27.1 and mean BMI of saline group is 27.7 with a p value of 0.450 which is insignificant.

Duration of surgery is also more with the saline group with a mean value of 125 minutes than the mean value of 119.5 in ketamine group with a p value of 0.436

Table 3: Body Mass Index and duration of surgery distribution of the study participants

Nebulisation		N	Mean	SD	P value
BMI	Ketamine	50	27.1	3.3	**0.450**
	Saline	50	27.7	4.0	
Duration of surgery (min)	Ketamine	50	119.5	35.4	**0.436**
	Saline	50	125.0	34.1	

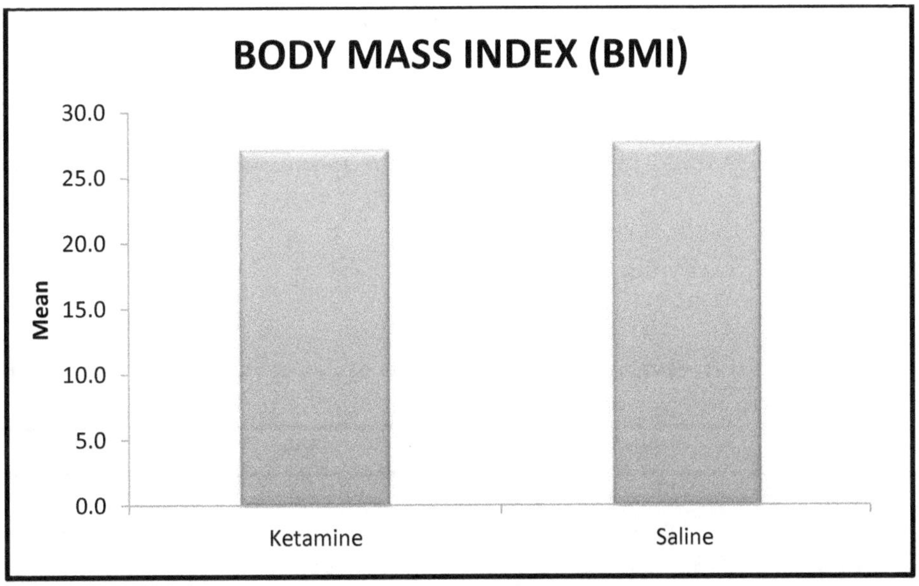

Fig 14: Body Mass Index distribution of the study participants

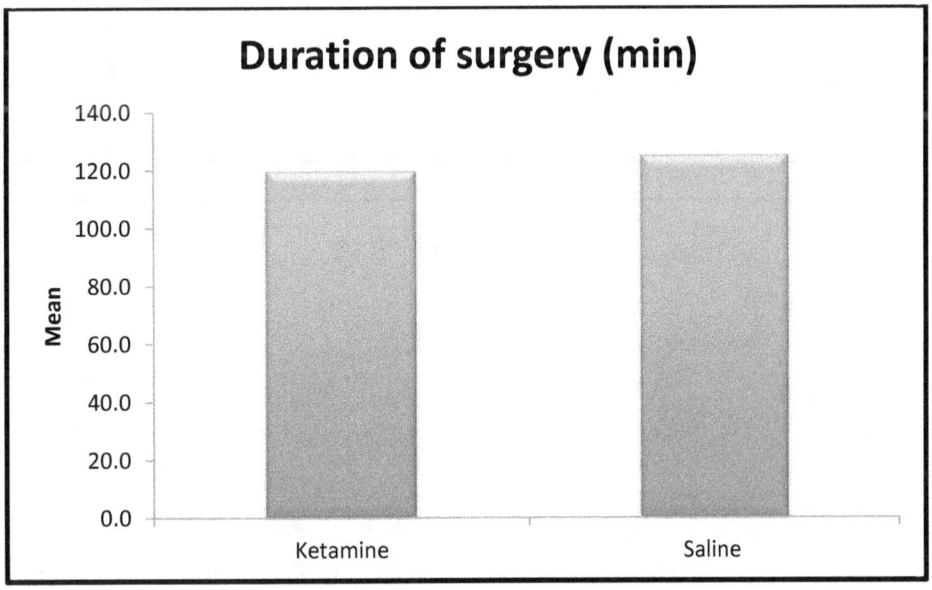

Figure 15: Duration of surgery distribution of the study participants

Heart rate:

During pre-induction period heart rate is found to be increased in ketamine group with a mean rate of 87.4/minute than in the mean heart rate of 80.6/minute in saline group with a significant p value of 0.024. whereas there is no much difference in heart rate between both the group thereafter.

Table 4: Distribution of heart rate among the study participants

HR	Nebulisation	N	Mean	SD	P value
Premed	Ketamine	50	83.2	11.5	**0.742**
	Saline	50	82.4	12.2	
PRE Induction	Ketamine	50	87.4	14.8	0.024
	Saline	50	80.6	14.7	
Induction	Ketamine	50	94.4	15.4	**0.932**
	Saline	50	94.1	14.9	
15 minutes	Ketamine	50	79.1	11.5	**0.792**
	Saline	50	78.5	11.2	
30 minutes	Ketamine	50	73.6	10.3	**0.487**
	Saline	50	75.0	10.1	
45 minutes	Ketamine	50	81.8	6.5	**0.681**
	Saline	50	81.3	5.5	

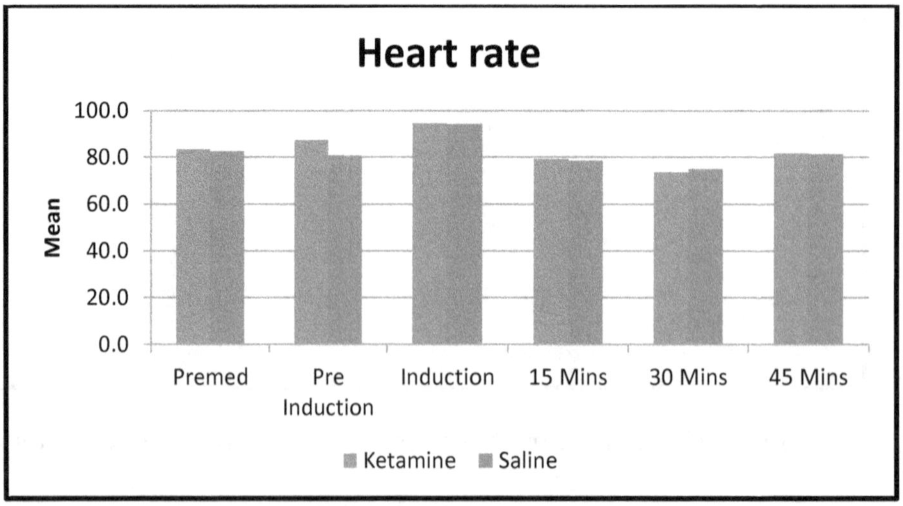

Fig 16: Changes in Heart rate among the study participants

SYSTOLIC BLOOD PRESSURE:

Systolic blood pressure in both the group were similar during pre-medication(basal), preinduction, induction, 15 minutes and 45 minutes whereas in 30 minutes after induction there is slight increase in systolic blood pressure in saline group of mean 119.3 than with the ketamine group of mean 121.2 with a significant p value of 0.035.

Table 5: Distribution of Systolic blood pressure among the study participants

SBP	Nebulisation	N	Mean	SD	P value
Premed	Ketamine	50	119.7	5.7	**0.256**
	Saline	50	121.0	5.8	
PRE Induction	Ketamine	50	115.3	8.5	**0.000**
	Saline	50	115.3	8.5	
Induction	Ketamine	50	128.5	4.8	**0.690**
	Saline	50	128.1	5.2	
15 minutes	Ketamine	50	106.8	10.0	**0.885**
	Saline	50	106.5	10.7	
30 minutes	Ketamine	50	119.3	4.5	0.035
	Saline	50	121.2	4.6	
45 minutes	Ketamine	50	125.4	3.4	**0.260**
	Saline	50	124.6	3.3	

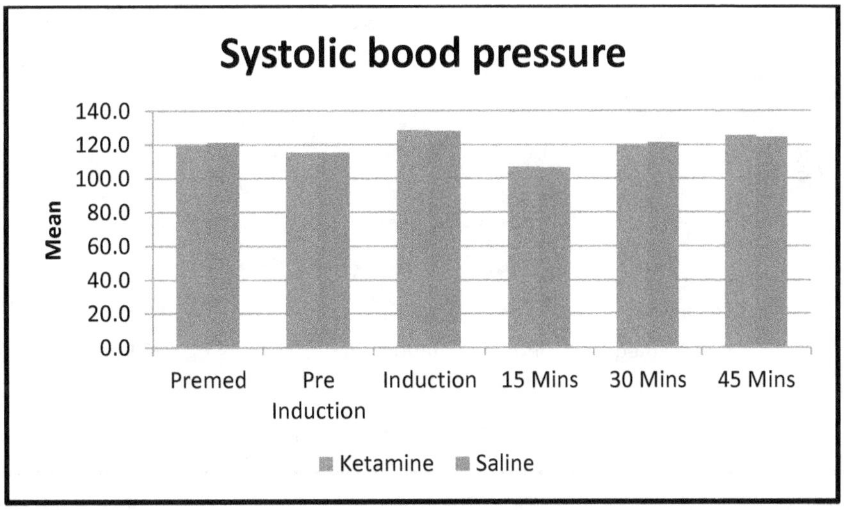

Fig 17: Changes in Systolic Blood Pressure among the study participants

DIASTOLIC BLOOD PRESSURE:

There is no statistical significance in diastolic blood pressure in both the groups

Table 6: Distribution of Diastolic blood pressure among the study groups

DBP	Nebulisation	N	Mean	SD	P value
Premed	Ketamine	50	76.9	7.2	**0.804**
	Saline	50	76.5	7.3	
PRE Induction	Ketamine	50	73.9	8.5	**0.692**
	Saline	50	74.6	7.6	
Induction	Ketamine	50	82.4	8.6	**0.098**
	Saline	50	85.3	8.7	
15 minutes	Ketamine	50	77.6	7.4	**0.989**
	Saline	50	77.6	7.1	
30 minutes	Ketamine	50	75.7	6.6	**0.828**
	Saline	50	75.4	7.2	
45 minutes	Ketamine	50	84.5	11.6	**0.836**
	Saline	50	84.0	10.5	

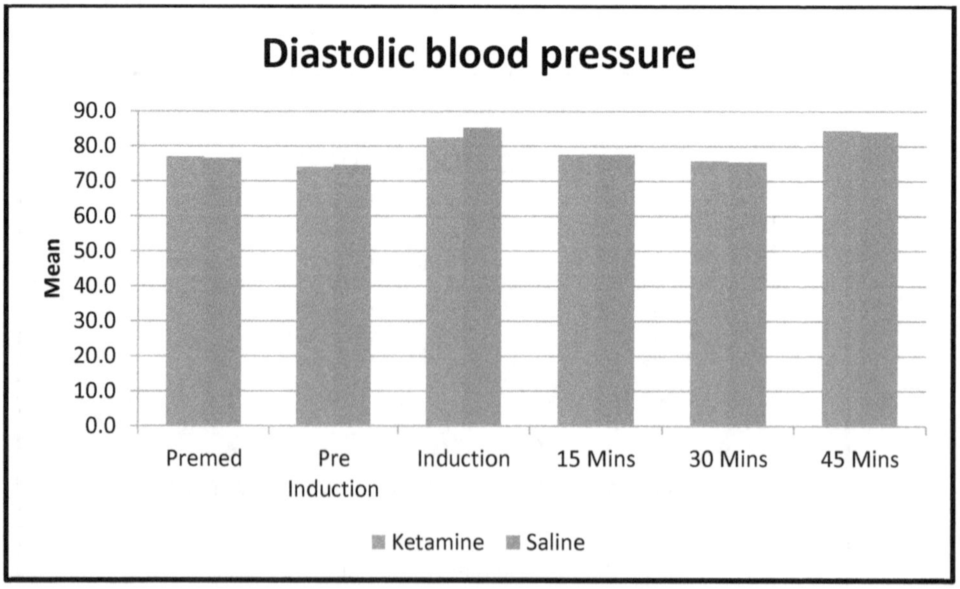

Fig 18: Changes in Diastolic blood pressure among the study participants

MEAN ARTERIAL PRESSURE:

Mean arterial pressure during premedication, pre induction, induction, 15 minutes, 30 minutes, 45 minutes following induction doesn't show any statistical significance.

Table 7: Distribution of Mean Arterial pressure among the study participants

MAP	Nebulisation	N	Mean	SD	P value
Premed	Ketamine	50	91.2	5.0	**0.842**
	Saline	50	91.4	5.0	
PRE Induction	Ketamine	50	87.7	6.4	**0.727**
	Saline	50	88.2	5.8	
Induction	Ketamine	50	97.8	6.2	**0.154**
	Saline	50	99.5	6.2	
15 minutes	Ketamine	50	87.3	6.1	**0.926**
	Saline	50	87.2	6.0	
30 minutes	Ketamine	50	90.2	4.8	**0.653**
	Saline	50	90.7	5.1	
45 minutes	Ketamine	50	98.1	8.0	**0.712**
	Saline	50	97.6	7.1	

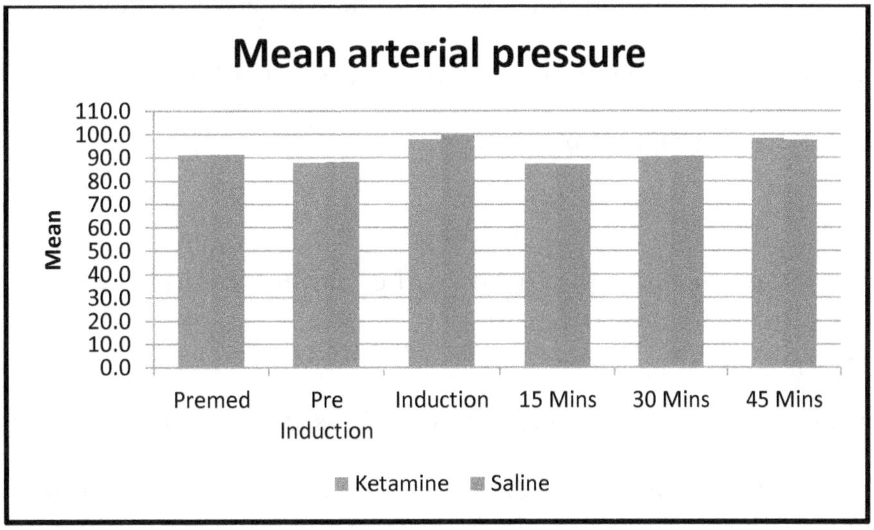

Fig 19: Changes in Mean arterial Pressure among the study participants

OXYGEN SATURATION:

There is no statistical difference in oxygen saturation of both the groups in different time interval.

Table 8: Distribution of Oxygen Saturation among the study participants

SPO2	Nebulisation	N	Mean	SD
Premed	Ketamine	50	98.9	0.9
	Saline	50	99.2	0.8
PRE Induction	Ketamine	50	99.5	0.5
	Saline	50	99.4	0.5
Induction	Ketamine	50	100.0	0.0
	Saline	50	100.0	0.0
15 minutes	Ketamine	50	100.0	0.0
	Saline	50	100.0	0.0
30 minutes	Ketamine	50	100.0	0.0
	Saline	50	100.0	0.0
45 minutes	Ketamine	50	100.0	0.0
	Saline	50	100.0	0.0

POST SCORE AFTER EXTUBATION (OT):

Out of 100 patients only 36 patients have a incidence of sore throat, whereas 64 patients have the score of 0 immediately after extubation in Operation theatre.

Out of this 64, 76% of the patient (n=38) belongs to the ketamine group than 56% of the patient (n=26) belongs to saline group. Hence the incidence of sore throat in immediate postoperative period were lesser with ketamine group with a significant p value of 0.037. Severe sore throat was seen with 2 patients in saline group and 0 in ketamine group.

Table 9: Distribution of POST SCORE AFTER EXTUBATION IN OT among the study participants

			Nebulisation		Total	P value
			Ketamine	Saline		
AFTER EXTUBATION POST SCORE(OT)	0	Count	38	26	64	0.037
		%	76.0%	52.0%	64.0%	
	1.0	Count	8	10	18	
		%	16.0%	20.0%	18.0%	
	2.0	Count	4	12	16	
		%	8.0%	24.0%	16.0%	
	3.0	Count	0	2	2	
		%	0.0%	4.0%	2.0%	
Total		Count	50	50	100	
		%	100.0%	100.0%	100.0%	

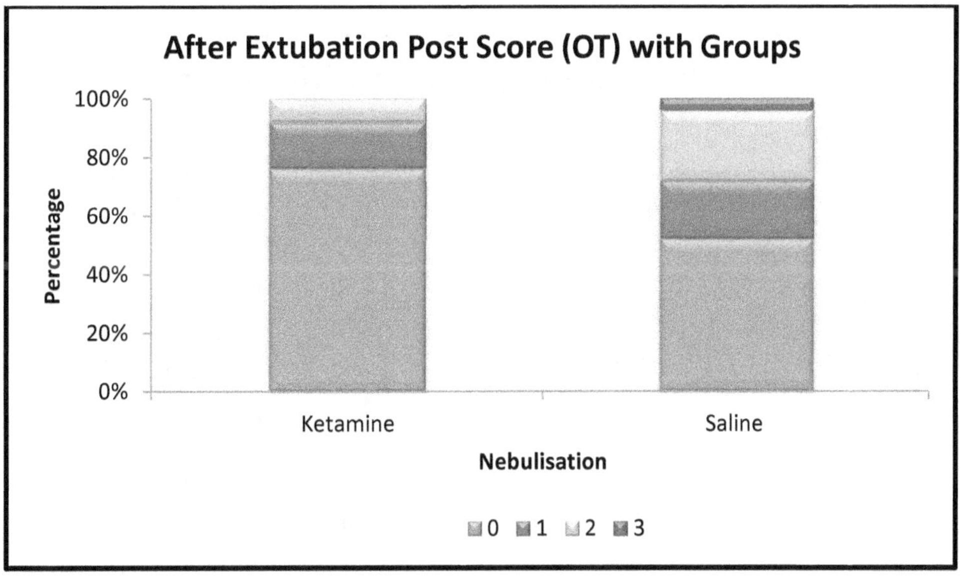

Fig 20: AFTER EXTUBATION IN OT POST SCORE distribution in study participants

POST SCORE RECOVERY ROOM:

Table 10: Distribution of RECOVERY ROOM POST SCORE "0" HOUR among the study participants

			Nebulisation		Total	P value
			Ketamine	Saline		
RECOVERY ROOM "O" HOUR	.0	Count	38	26	64	0.037
		%	76.0%	52.0%	64.0%	
	1.0	Count	8	10	18	
		%	16.0%	20.0%	18.0%	
	2.0	Count	4	12	16	
		%	8.0%	24.0%	16.0%	
	3.0	Count	0	2	2	
		%	0.0%	4.0%	2.0%	
Total		Count	50	50	100	
		%	100.0%	100.0%	100.0%	

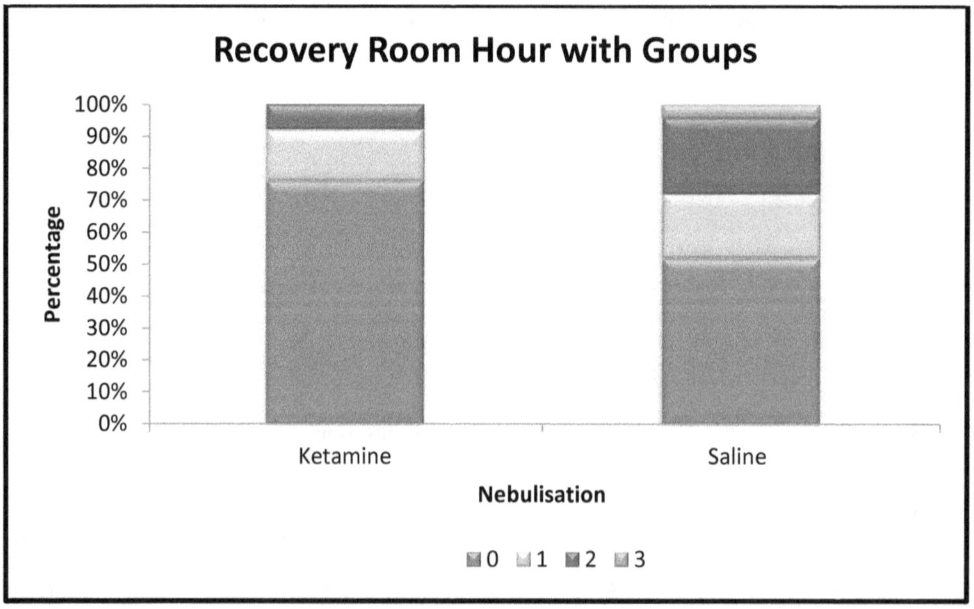

Fig 21: RECOVERY ROOM POST SCORE "0" HOUR distribution in study participants

POST SCORE 2ND HOUR:

Table 11: Distribution of POST SCORE "2nd" HOUR among the study participants

			Nebulisation		Total	P value
			Ketamine	Saline		
2ND HOUR	.0	Count	42	26	68	0.005
		%	84.0%	52.0%	68.0%	
	1.0	Count	5	10	15	
		%	10.0%	20.0%	15.0%	
	2.0	Count	3	12	15	
		%	6.0%	24.0%	15.0%	
	3.0	Count	0	2	2	
		%	0.0%	4.0%	2.0%	
Total		Count	50	50	100	
		%	100.0%	100.0%	100.0%	

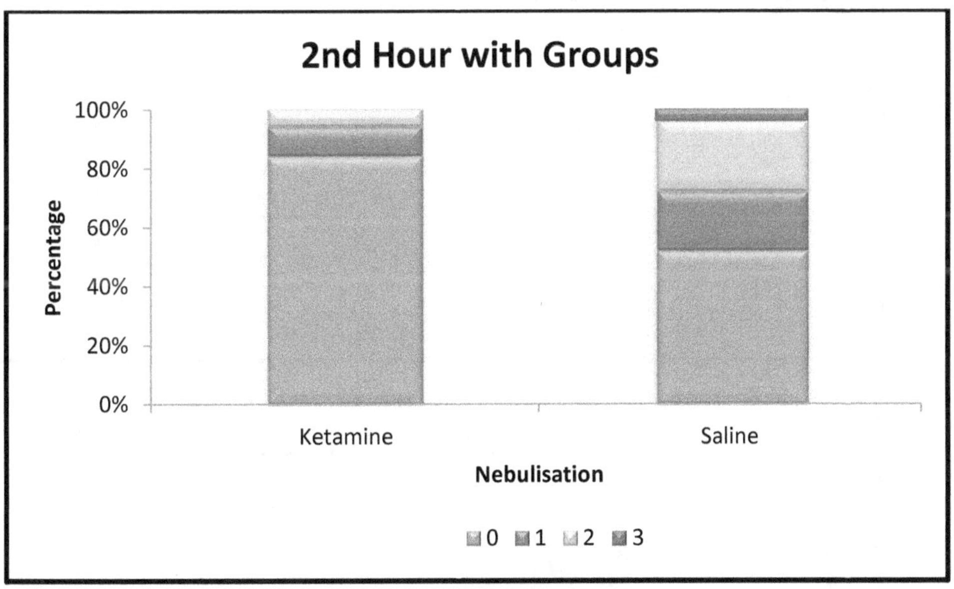

Fig 22: POST SCORE "2nd" HOUR distribution in study participants

POST SCORE 4TH HOUR:

During 4th hour number of patient having no sore throat improved to 42 patients (84.0%) in ketamine group and 28 patient in saline group(56.0%) having a significant p value of 0.013

Whereas the patient having severe sore throat doesn't improve and have the same distribution of 2 patient in saline group.

Table 12: Distribution of POST SCORE "4th" HOUR among the study participants

			Nebulisation		Total	P value
			Ketamine	Saline		
4TH HOUR	.0	Count	42	28	70	0.013
		%	84.0%	56.0%	70.0%	
	1.0	Count	6	11	17	
		%	12.0%	22.0%	17.0%	
	2.0	Count	2	9	11	
		%	4.0%	18.0%	11.0%	
	3.0	Count	0	2	2	
		%	0.0%	4.0%	2.0%	
Total		Count	50	50	100	
		%	100.0%	100.0%	100.0%	

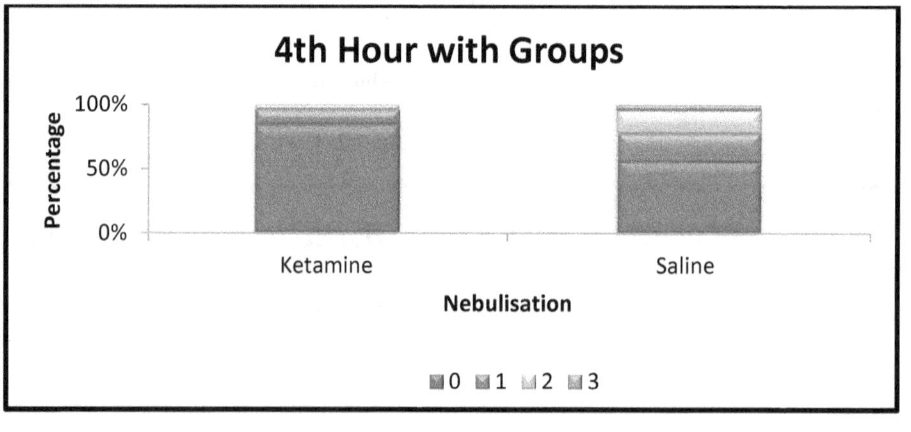

Fig 23: POST SCORE "4th" HOUR distribution in study participants

POST SCORE 8TH HOUR:

During 8th hour POST score is found to be 1 (mild) in ketamine group for 6 (12%) patients and 12 (24%) patients in saline group and no patient have POST score of 2(moderate), 3(severe) in ketamine group having a significant p value of 0.005. Severe sore throat was seen in saline group for one patient alone.

Table 13: Distribution of POST SCORE "8th" HOUR among the study participants

			Nebulisation		Total	P value
			Ketamine	Saline		
8TH HOUR	.0	Count	44	30	74	0.005
		%	88.0%	60.0%	74.0%	
	1.0	Count	6	12	18	
		%	12.0%	24.0%	18.0%	
	2.0	Count	0	7	7	
		%	0.0%	14.0%	7.0%	
	3.0	Count	0	1	1	
		%	0.0%	2.0%	1.0%	
Total		Count	50	50	100	
		%	100.0%	100.0%	100.0%	

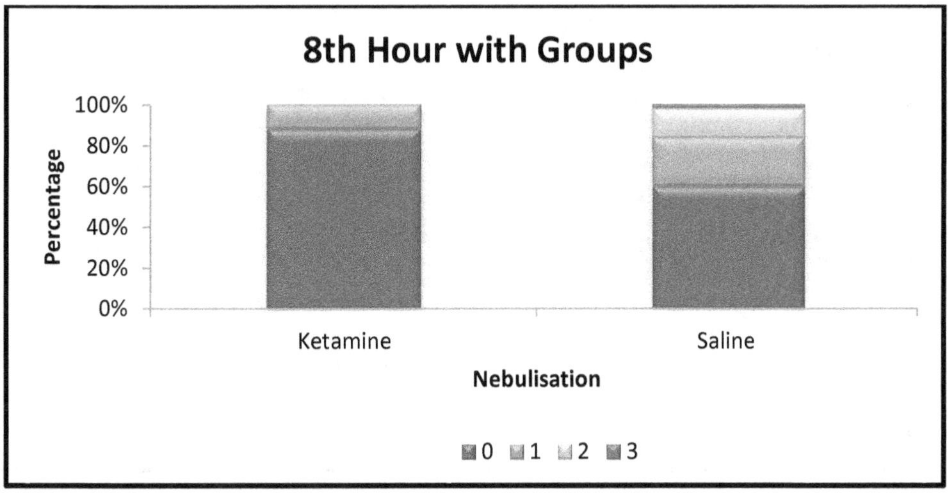

Fig 24: POST SCORE "8th" HOUR distribution in study participants

POST SCORE 12TH HOUR:

In the period of 12th hour no severe sorethroat (POST VALUE of 0) were found in both the groups and patient with no sorethroat also improved to 47 patients (94.0%) in ketamine group and 35 patients (70.0%) in saline group with a significant p value of 0.006.

Table 14: Distribution of POST SCORE "12th" HOUR among the study participants

			Nebulisation		Total	P value
			Ketamine	Saline		
12TH HOUR	.0	Count	47	35	82	0.006
		%	94.0%	70.0%	82.0%	
	1.0	Count	3	11	14	
		%	6.0%	22.0%	14.0%	
	2.0	Count	0	4	4	
		%	0.0%	8.0%	4.0%	
Total		Count	50	50	100	
		%	100.0%	100.0%	100.0%	

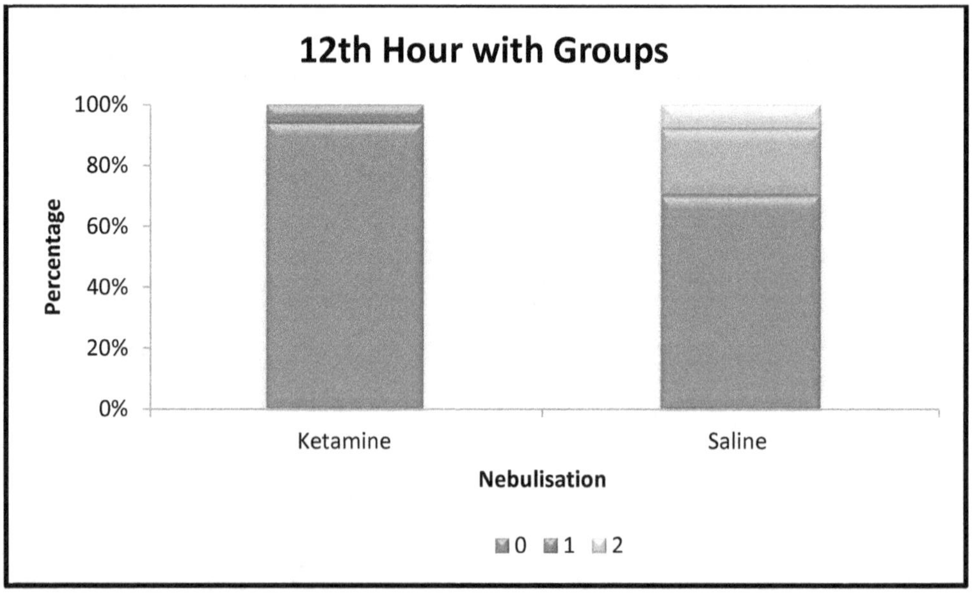

Fig 25: POST SCORE "12th" HOUR distribution in study participants

POST SCORE 24TH HOUR:

Even at 24th hour moderate sore throat was found in one patient, mild sorethroat in 8 patients in saline group whereas no patient with moderate sore throat and one patient with mild sorethroat in ketamine group.

At the end of the 24th hour 98% of the patient have no sorethroat in ketamine group and 82% of the patients have no sore throat in saline group with a siginificant p value of 0.028.

Table 15: Distribution of POST SCORE "24th" HOUR among the study participants

			Nebulisation		Total	P value
			Ketamine	Saline		
24TH HOUR	.0	Count	49	41	90	0.028
		%	98.0%	82.0%	90.0%	
	1.0	Count	1	8	9	
		%	2.0%	16.0%	9.0%	
	2.0	Count	0	1	1	
		%	0.0%	2.0%	1.0%	
Total		Count	50	50	100	
		%	100.0%	100.0%	100.0%	

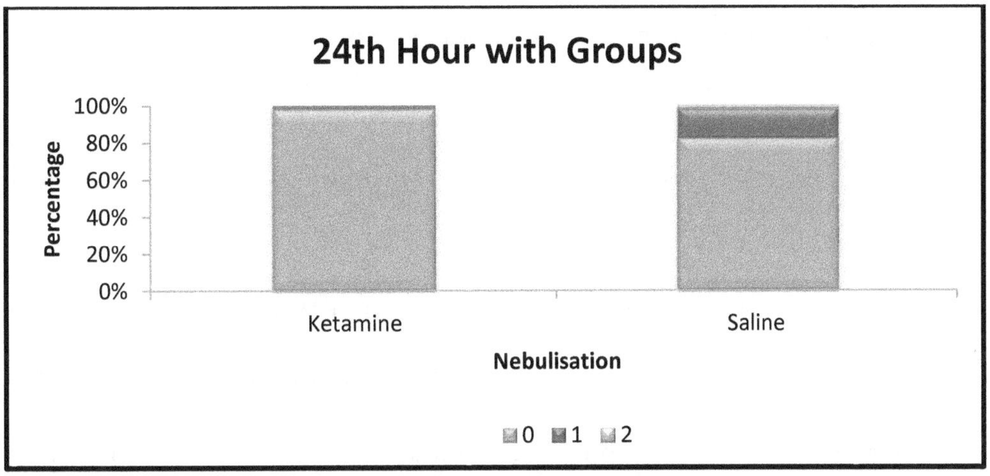

Fig 26: POST SCORE "24th" HOUR distribution in study participants

DISCUSSION

Postoperative sore throat is the most common complaint of patients requiring endotracheal intubation following general anaesthesia. Postoperative sore throat persists despite the numerous studies and evaluations aimed at reducing its incidence and severity. It is found to be more prevalent during lengthy surgical procedures.

In patients with post-operative sore throat, minor symptoms such as irritation of the throat, difficulty in swallowing, hoarseness of voice, and even laryngitis and tracheitis are possible. Multiple factors including mucosal tear during laryngoscopy and intubation, pressure exerted on the tracheal wall by an inflated cuff, and dehydration of tracheal mucosa, contribute to post-operative sore throat. Age, sex, the use of succinylcholine, and cuff pressure are additional contributors to POST. Utilizing a multimodal approach consisting of non-pharmacological and pharmacological interventions can reduce postoperative sore throat.

Study done by Bhagyashree Amingad et al[38] on comparison of gargle and nebulisation of ketamine for reduction of postoperative sore throat concludes that both the form of drug deposition found to be efficient in reducing the severity of post operative sore throat, but in our study we have used nebulisation as a source of drug delivery for the following reason such as need of lesser volume of drug and reduced chances of aspiration

All demographic parameter like age, sex, height, weight, BMI, ASA physical status and mean duration of surgery were comparable between both groups and statistically insignificant.

In this present study heart rate following ketamine nebulisation is found to be increased with a mean heart rate of 86.4/minute versus 80.6/minute in saline group which can be due to sympathomimetic effect of ketamine.

Similarly systolic blood pressure following 30 minutes after induction is found to be increased in saline group with mean SBP of 121.2 mmHg versus 119.3mmHg in ketamine group. This finding can be due to analgesic effect of the ketamine.

The incidence of sore throat was evaluated immediately following extubation in the operating room, as well as at 0hr, 2hr, 4hr, 12hr, and 24hr in the PACU. At 0 hour ($p=0.037$), 2nd hour ($p=0.005$), 4th hour ($p=0.013$), 8th hour ($p=0.005$), 12th hour ($p=0.006$), and 24th hour ($p=0.028$), a statistically significant reduction in sore throat was observed in the ketamine group compared to the saline group.

Shrestha s et al[39] determine the incidence and associated risk factors of postoperative sore throat in which the incidence of postoperative sore throat was found to be 50.8% in endotracheal tube intubation and it is associated more in female patients and in older age group, whereas in our study the incidence of postoperative sore throat was 36% overall and most common age group in our study is from 40-60 years of age, but we didn't compare the incidence of sore throat between the genders and in different age groups.

Similarly there is a Study by rudra et al[41] compared the effectiveness of ketamine gargles with placebo in preventing post operative sore throat after endotracheal intubation in 40 patients in which the result shows incidence of POST in saline group is 85% at 4th hour which is reduced up to 60% in 24th hours whereas in ketamine group 40% in 4th hour which is reduced up to 25% at 24th hour, but in our study of 100 patients results shows POST in saline group is 44% at 4th hour which got reduced to 18% in 24th hour whereas in ketamine group incidence of sore throat is only 16% at 4th hour which got reduced to only 2% in 24th hour. Due to more sample size in our study there might be increased incidence but the severity of sore throat is significantly reduced in our study which is probably due to the effective drug deposition through nebulisation in our study whereas it is gargle in the study which we compared.

Comparison between ketamine 50mg nebulisation and different doses such as 250 and 500mg of magnesium sulphate nebulisation was done by Rajan et al[40] concludes that it reduces the incidence and severity of Post operative sore throat at 0, 2nd and 4th hours in both ketamine and 500mg magnesium sulphate group, which is found to be similar to our study which also shows a reduction in incidence of sore throat at 0, 2nd and 4th hour with a significant reduction in the severity also.

Localized trauma resulting in aseptic inflammation of the pharyngeal mucosa could be the cause of orotracheal tube-related sore throat. It may also cause oedema, congestion, and pain [41,42]

Reduced expression of inducible nitric oxide synthase and attenuation in central component of inflammatory changes have found to be reason for the anti-inflammatory properties of ketamine.

We used nebulized form of ketamine in this study instead of its other forms such as oral, gargle or intravenous due to the fact that it can be easily administered before the surgery and found to be effective form of deposition. In our study we used a wall-mounted oxygen driven nebulization method. In this method, compressed air will break down the liquids into droplets. Larger particles of 10–25 μm get deposited in mouth and throat and particle with 5–10 μm diameter deposited in a passage from mouth to airway.[43] The settling of aerosol in mouth and upper airway might probably be the reason for the decreased incidence and severity of POST in ketamine group, due to its topical analgesic, anti-inflammatory, and NMDA-receptor antagonistic effect.

M reddy et al[22] studied dose-dependent effectiveness of ketamine nebulisation in preventing post-operative sore throat due to tracheal intubation and result shows that nebulized ketamine at a dose of 0.5mg/kg was comparatively less effective than 1 mg/kg and 1.5 mg/kg. He found out that the increased and thicker secretions in patients who received 1.5 mg/kg dose of ketamine nebulization. In our study we have used standard dose of 25mg in 4.5ml of normal saline irrespective of patient weight to avoid side effects of ketamine such as sedation or increased secretions. The serum ketamine levels were also not monitored as small dosage were used.

Patient usually receives paracetamol in postoperative period. This rescue analgesia will not affect subsequent assessment of sore throat because a study by Mishra J et al[44] has shown that intravenous paracetamol decreased incidence of sore throat for up to 2hrs only.

There was no incidence of any systemic toxicity, or any other adverse reaction seen in post operative period in any study group.

LIMITATIONS

1. The first limitation being dose we used was a fixed one irrespective of the weight of the patient or gender and hence the least effective dose is not known.
2. Second limitation such as laryngoscopy was done by different anaesthesiologist and factors which affects the result of POST such as number of attempts, time taken for tracheal intubation, other airway instrumentation such as oropharyngeal airway, Ryle's tube or presence of nasopharyngeal airway were not taken into considerations.
3. We have not measured the serum levels of ketamine and its metabolite nor ketamine to rule out systemic effects of the same
4. Apart from all these limitations main limitation being a subjective variation of assessing a severity of sore throat is always there in our study.

CONCLUSION

In comparison to normal saline nebulization, preoperative nebulization with Ketamine (25 mg) significantly reduces the incidence and severity of postoperative sore throat.

SUMMARY

TITLE: A clinical study to evaluate the efficacy of nebulised ketamine over normal saline prior to general anaesthesia in reducing the incidence and severity of post operative sore throat.

Aim of the research project:

The aim of the study is to evaluate the efficacy of ketamine nebulisation versus normal saline given 15 min before general anaesthesia in preventing the post-operative sore throat and to assess the POST severity score in ASA 1 and 2 patients aged between 18 to 65 years undergoing elective surgeries.

Methodology:

All patients were instructed to fast for at least eight hours prior to surgery. Each patient was given an explanation of general anaesthesia (GA) and our study, and written informed consent was obtained from the patient and his family in the presence of an independent witness. The investigators on the surgical ward enrolled patients the day prior to surgery.

Using opaque, sealed envelopes containing computer-generated random number tables, a non-participating anaesthesiologists randomised patients into two groups. The staff nurse opened the envelopes and prepared nebulization solution according to group allocation. Patients in Group A received ketamine (4.5ml of normal saline and 0.5ml of ketamine 25mg) while patients in Group B received 5ml of normal saline. Patients received the study drug via nebulization mask connected to an oxygen-driven wall-mounted source (8 L) for 15 minutes prior to surgery. The registered nurse did not take part in the subsequent evaluation of these patients.

At pre-nebulization (baseline parameters), pre-induction, after induction, 15 minutes, 30 minutes, 45 minutes, and 60 minutes of GA, hemodynamic (mean BP, mean pulse rate, and SpO2) measurements were obtained. The severity of sore throat was evaluated at 0 h, 2 h, 4 h, 8 h, 12 h, and 24 h postoperatively

RESULTS:

Out of 100 patients more than 40% of the age group in our study falls under 40-60 years. There are no statistically significant differences between the group with respect to age (p value-0.289)

Gender of the study participants are taken equally in ketamine groups whereas male gender predominates in saline group hence in total male patients are more than the female patients with a p value of 0.315. Body mass index in each group is almost similar with the mean BMI of ketamine group is 27.1 and mean BMI of saline group is 27.7 with a p value of 0.450 which is insignificant.

Duration of surgery is also more with the saline group with a mean value of 125 minutes than the mean value of 119.5 in ketamine group with a p value of 0.436

There is no much difference in baseline heart rate between the groups, but during pre-induction period mean heart is found to be increased in ketamine group with a mean rate of 87.4 than in the mean heart rate of 80.6 in saline group with a significant p value of 0.024. whereas there is no much difference in heart rate between both the group during induction, 15, 30 and 45 minutes.

Systolic blood pressure in both the group were similar during pre-medication, preinduction, induction, 15 minutes and 45 minutes whereas in 30 minutes after induction there is slight increase in systolic blood pressure in saline group of mean 119.3 than with the ketamine group of mean 121.2 with a significant p value of 0.035. There is no statistically significant difference between diastolic blood pressure, mean arterial pressure and oxygen saturation.

Out of 100 patients only 36 patients have a incidence of sore throat, whereas 64 patients have the score of 0 immediately after extubation in Operation theatre.

Out of this 64, 76% of the patient (n=38) belongs to the ketamine group than 56% of the patient (n=26) belongs to saline group. Hence the incidence of sore throat in immediate postoperative period were lesser with ketamine group with a significant p value of 0.037. Severe sore throat was seen with 2 patients in saline group and 0 in ketamine group.

During 4th hour number of patient having no sore throat improved to 42 patients (84.0%) in ketamine group and 28 patient in saline group (56.0%) having a significant p value of 0.013, whereas the 2 patients having severe sore throat doesn't improve in saline group.

In the period of 12th hour, no severe sore throat (POST VALUE of 0) were found in both the groups and patient with no sore throat also improved to 47 patients (94.0%) in ketamine group and 35 patients (70.0%) in saline group with a significant p value of 0.006.

Even at 24th hour moderate sore throat was found in one patient, mild sore throat in 8 patients in saline group whereas no patient with moderate sore throat and one patient with mild sore throat in ketamine group.

At the end of the 24th hour 98% of the patient have no sore throat in ketamine group and 82% of the patients have no sore throat in saline group with a significant p value of 0.028.

REFERENCES

1. Higgins, P. P., Chung, F., & Mezei, G. (2002). Postoperative sore throat after ambulatory surgery. *British Journal of Anaesthesia*, *88*(4), 582–584. https://doi.org/10.1093/bja/88.4.582
2. Loeser, E. A., Bennett, G. M., Orr, D. L., & Stanley, T. H. (1980). Reduction of postoperative sore throat with new endotracheal tube cuffs. *Anesthesiology*, *52*(3), 257–259. https://doi.org/10.1097/00000542-198003000-00011
3. Macario, A., Weinger, M., Carney, S., & Kim, A. (1999). Which clinical anesthesia outcomes are important to avoid? The perspective of patients. *Anesthesia and Analgesia*, *89*(3), 652–658. https://doi.org/10.1097/00000539-199909000-00022
4. Sun, L., Guo, R., & Sun, L. (2014). Dexamethasone for preventing postoperative sore throat: A meta-analysis of randomized controlled trials. *Irish Journal of Medical Science*, *183*(4), 593–600. https://doi.org/10.1007/s11845-013-1057-0
5. McHardy, F. E., & Chung, F. (1999). Postoperative sore throat: Cause, prevention and treatment. *Anaesthesia*, *54*(5), 444–453. https://doi.org/10.1046/j.1365-2044.1999.00780.x
6. Al-Qahtani, A. S., & Messahel, F. M. (2005). Quality improvement in anesthetic practice—Incidence of sore throat

after using small tracheal tube. *Middle East Journal of Anaesthesiology*, *18*(1), 179–183.

7. Suzuki, N., Kooguchi, K., Mizobe, T., Hirose, M., Takano, Y., & Tanaka, Y. (1999). [Postoperative hoarseness and sore throat after tracheal intubation: Effect of a low intracuff pressure of endotracheal tube and the usefulness of cuff pressure indicator]. *Masui. The Japanese Journal of Anesthesiology*, *48*(10), 1091–1095.
8. Estebe, J.-P., Delahaye, S., Le Corre, P., Dollo, G., Le Naoures, A., Chevanne, F., & Ecoffey, C. (2004). Alkalinization of intra-cuff lidocaine and use of gel lubrication protect against tracheal tube-induced emergence phenomena. *British Journal of Anaesthesia*, *92*(3), 361–366. https://doi.org/10.1093/bja/aeh078
9. Jung, H., & Kim, H. J. (2013). Dexamethasone contributes to the patient management after ambulatory laryngeal microsurgery by reducing sore throat. *European Archives of Oto-Rhino-Laryngology: Official Journal of the European Federation of Oto-Rhino-Laryngological Societies (EUFOS): Affiliated with the German Society for Oto-Rhino-Laryngology - Head and Neck Surgery*, *270*(12), 3115–3119. https://doi.org/10.1007/s00405-013-2708-5
10. Sumathi, P. A., Shenoy, T., Ambareesha, M., & Krishna, H. M. (2008). Controlled comparison between betamethasone gel and lidocaine jelly applied over tracheal tube to reduce

postoperative sore throat, cough, and hoarseness of voice. *British Journal of Anaesthesia*, *100*(2), 215–218. https://doi.org/10.1093/bja/aem341

11. Tazeh-Kand, N. F., Eslami, B., & Mohammadian, K. (2010). Inhaled fluticasone propionate reduces postoperative sore throat, cough, and hoarseness. *Anesthesia and Analgesia*, *111*(4), 895–898. https://doi.org/10.1213/ANE.0b013e3181c8a5a2
12. Lévy, B., Mouillac, F., Quilichini, D., Schmitz, J., Gaudart, J., & Gouin, F. (2003). [Topical methylprednisolone vs lidocaïne for the prevention of postoperative sore throat]. *Annales Francaises D'anesthesie Et De Reanimation*, *22*(7), 595–599. https://doi.org/10.1016/s0750-7658(03)00214-4
13. Jiao, L., & Liu, R. (2014). [Effects of dezocine on postoperative sore throat after maxillofacial procedures: A comparison with flurbiprofen axetil]. *Beijing Da Xue Xue Bao. Yi Xue Ban = Journal of Peking University. Health Sciences*, *46*(1), 104–106.
14. Agarwal, A., Nath, S. S., Goswami, D., Gupta, D., Dhiraaj, S., & Singh, P. K. (2006). An evaluation of the efficacy of aspirin and benzydamine hydrochloride gargle for attenuating postoperative sore throat: A prospective, randomized, single-blind study. *Anesthesia and Analgesia*, *103*(4), 1001–1003. https://doi.org/10.1213/01.ane.0000231637.28427.00

15. Maruyama, K., Yamada, T., & Hara, K. (2006). Effect of clonidine premedication on postoperative sore throat and hoarseness after total intravenous anesthesia. *Journal of Anesthesia*, *20*(4), 327–330. https://doi.org/10.1007/s00540-006-0421-x

16. Ahuja, V., Mitra, S., & Sarna, R. (2015). Nebulized ketamine decreases incidence and severity of post-operative sore throat. *Indian Journal of Anaesthesia*, *59*(1), 37–42. https://doi.org/10.4103/0019-5049.149448

17. Borazan, H., Kececioglu, A., Okesli, S., & Otelcioglu, S. (2012). Oral magnesium lozenge reduces postoperative sore throat: A randomized, prospective, placebo-controlled study. *Anesthesiology*, *117*(3), 512–518. https://doi.org/10.1097/ALN.0b013e3182639d5f

18. Canbay, O., Celebi, N., Sahin, A., Celiker, V., Ozgen, S., & Aypar, U. (2008). Ketamine gargle for attenuating postoperative sore throat. *British Journal of Anaesthesia*, *100*(4), 490–493. https://doi.org/10.1093/bja/aen023

19. Zhu, M. M., Zhou, Q. H., Zhu, M. H., Rong, H. B., Xu, Y. M., Qian, Y. N., & Fu, C. Z. (2007). Effects of nebulized ketamine on allergen-induced airway hyperresponsiveness and inflammation in actively sensitized Brown-Norway rats. *Journal of Inflammation (London, England)*, *4*, 10. https://doi.org/10.1186/1476-9255-4-10

20. Dhuliya, S., Patel, N., & Shah, D. (2022). Comparative evaluation of incidence of post operative sore throat after nebulization with ketamine and magnesium sulfate in patients undergoing general anaesthesia requiring endotracheal intubation. *Indian Journal of Clinical Anaesthesia*, *9*(2), 227–232. https://doi.org/10.18231/j.ijca.2022.045

21. Mostafa, R. H., Saleh, A. N., & Hussein, M. M. (2018). A Comparative Study of Three Nebulized Medications for the Prevention of Postoperative Sore Throat in the Pediatric Population. *The Open Anesthesia Journal*, *12*(1), 85–93. https://doi.org/10.2174/2589645801812010085

22. Reddy, M., & Fiaz, S. (2017). Dose-dependent effectiveness of ketamine nebulisation in preventing post-operative sore throat due to tracheal intubation. *Sri Lankan Journal of Anaesthesiology*, *26*(1), 22. https://doi.org/10.4038/slja.v26i1.8264

23. Shroff, P. P., & Patil, V. (2009). Efficacy of cuff inflation media to prevent postintubation-related emergence phenomenon: Air, saline and alkalinized lignocaine: *European Journal of Anaesthesiology*, *26*(6), 458–462. https://doi.org/10.1097/EJA.0b013e32832403fa

24. *(PDF) Effects of Endotracheal Tube Size and Cuff Pressure on the Incidence of Postoperative Sore Throat: Comparison Between Three Facilities*. (n.d.). Retrieved November 17, 2022, from

https://www.researchgate.net/publication/338213993_Effects_of_Endotracheal_Tube_Size_and_Cuff_Pressure_on_the_Incidence_of_Postoperative_Sore_Throat_Comparison_Between_Three_Facilities

25. Fenta, E., Teshome, D., Melaku, D., & Tesfaw, A. (2020). Incidence and factors associated with postoperative sore throat for patients undergoing surgery under general anesthesia with endotracheal intubation at Debre Tabor General Hospital, North central Ethiopia: A cross-sectional study. *International Journal of Surgery Open*, *25*, 1–5. https://doi.org/10.1016/j.ijso.2020.06.003

26. Su, Z., Li, S., Zhou, Z., Chen, X., Gu, Y., Chen, Y., Zhong, C., Zhong, M., & Zhong, N. (2017). A canine model of tracheal stenosis induced by cuffed endotracheal intubation. *Scientific Reports*, *7*(1), 45357. https://doi.org/10.1038/srep45357

27. Joffe, A. M., Aziz, M. F., Posner, K. L., Duggan, L. V., Mincer, S. L., & Domino, K. B. (2019). Management of Difficult Tracheal Intubation. *Anesthesiology*, *131*(4), 818–829. https://doi.org/10.1097/ALN.0000000000002815

28. Agro, F., Frass, M., Benumof, J. L., & Krafft, P. (2002). Current status of the Combitube™: A review of the literature. *Journal of Clinical Anesthesia*, *14*(4), 307–314. https://doi.org/10.1016/S0952-8180(02)00356-2

29. Wissler, R. N. (1993). The esophageal-tracheal Combitube. *Anesthesiology Review*, *20*(4), 147–152.

30. Rich, J. M., Mason, A. M., Bey, T. A., Krafft, P., & Frass, M. (2004). The critical airway, rescue ventilation, and the combitube: Part 2. *AANA Journal*, *72*(2), 115–124.

31. Urtubia, R. M., Aguila, C. M., & Cumsille, M. A. (2000). Combitube: A study for proper use. *Anesthesia and Analgesia*, *90*(4), 958–962. https://doi.org/10.1097/00000539-200004000-00034

32. Walz, R., Davis, S., & Panning, B. (1999). Is the Combitube a useful emergency airway device for anesthesiologists? *Anesthesia and Analgesia*, *88*(1), 233. https://doi.org/10.1097/00000539-199901000-00056

33. Gaitini, L. A., Vaida, S. J., & Agro, F. (2002). The Esophageal-Tracheal Combitube. *Anesthesiology Clinics of North America*, *20*(4), 893–906. https://doi.org/10.1016/s0889-8537(02)00021-4

34. Cates, C. J., Welsh, E. J., & Rowe, B. H. (2013). Holding chambers (spacers) versus nebulisers for beta-agonist treatment of acute asthma. *Cochrane Database of Systematic Reviews*. https://doi.org/10.1002/14651858.CD000052.pub3

35. Kono, K., Watanabe, T., Dote, T., Usuda, K., Nishiura, H., Tagawa, T., Tominaga, M., Higuchi, Y., & Onnda, M. (2000). Successful treatments of lung injury and skin burn due to hydrofluoric acid exposure. *International Archives of*

Occupational and Environmental Health, *73 Suppl*, S93-97. https://doi.org/10.1007/pl00014634

36. Hickey, A. J. (Ed.). (2004). *Pharmaceutical inhalation aerosol technology* (2nd ed., rev.expanded). M. Dekker.
37. Jendle, J., Karlberg, B. E., Persliden, J., Franzén, L., & Arborelius, M. (1995). Delivery and retention of an insulin aerosol produced by a new jet nebulizer. *Journal of Aerosol Medicine: The Official Journal of the International Society for Aerosols in Medicine*, *8*(3), 243–254. https://doi.org/10.1089/jam.1995.8.243
38. Amingad, B., & Jayaram, S. (2016). Comparison of ketamine nebulisation with ketamine gargle in attenuating post operative sore throat. *Indian Journal of Clinical Anaesthesia*, *3*(3), 347. https://doi.org/10.5958/2394-4994.2016.00060.3
39. Shrestha, S., Maharjan, B., & Karmacharya, R. M. (2017). Incidence and Associated Risk Factors of Postoperative Sore Throat in Tertiary Care Hospital. *Kathmandu University Medical Journal (KUMJ)*, *15*(57), 10–13.
40. Rajan, S., Malayil, G., Varghese, R., & Kumar, L. (2017). Comparison of usefulness of ketamine and magnesium sulfate nebulizations for attenuating postoperative sore throat, hoarseness of voice, and cough. *Anesthesia: Essays and Researches*, *11*(2), 287. https://doi.org/10.4103/0259-1162.181427

41. Elhakim, M., Siam, A., Rashed, I., & Hamdy, M. H. (2000). Topical tenoxicam from pharyngeal pack reduces postoperative sore throat: Intraoperative tenoxicam pharyngeal pack. *Acta Anaesthesiologica Scandinavica*, *44*(6), 733–736. https://doi.org/10.1034/j.1399-6576.2000.440612.x

42. Neder Meyer, T., & Lázaro da Silva, A. (2004). Ketamine reduces mortality of severely burnt rats, when compared to midazolam plus fentanyl. *Burns*, *30*(5), 425–430. https://doi.org/10.1016/j.burns.2004.01.006

43. O'Callaghan, C., & Barry, P. W. (1997). The science of nebulised drug delivery. *Thorax*, *52*(Supplement 2), S31–S44. https://doi.org/10.1136/thx.52.2008.S31

44. Mishra, J., Nanda, S., Dalai, H., Das, K., Pattanaik, B., & Surbhi, S. (2017). PARACETAMOL IN PREVENTION OF POSTOPERATIVE SORE THROAT- AN OBSERVATIONAL STUDY. *Journal of Evidence Based Medicine and Healthcare*, *4*(57), 3434–3437. https://doi.org/10.18410/jebmh/2017/684

ANNEXURE

ANNEXURE I

CASE PROFORMA:

NAME: AGE: GENDER: I.D NUMBER:

HEIGHT: WEIGHT: BMI:

ASA CLASSIFICATION: DIAGNOSIS: PROCEDURE:

TIME OF NEBULISATION: TIME OF INDUCTION: STARTING OF SURGERY: ENDING OF SURGERY:

VITALS CHART:

	TIME	HR	SPO2	SBP	DBP	MAP
PRE NEBULISATION						
PRE INDUCTION						
INDUCTION						
15MINUTES						
30MINUTES						
45MINUTES						
60MINUTES						

SORE THROAT ASSESSMENT:

	TIME	SCORE
AFTER EXTUBATION AT OT		
RECOVERY ROOM (0 HOUR)		
2nd HOUR		
4TH HOUR		
8th HOUR		
12th HOUR		
24th HOUR		

SORE THROAT ASSESSMENT SCORE:

SCORE 0= NO SORE THROAT

SCORE 1=MILD SORE THROAT (complains of sore throat only on asking)

SCORE 2=MODERATE SORE THROAT (complains of sore throat his/her own)

SCORE 3=SEVERE SORE THROAT (change of voice or hoarseness, associated with throat pain)

OTHER COMPLICATIONS: (if anything)

ANNEXURE II

PARTICIPANT/PATIENT INFORMATION SHEET

Title of the research project: A clinical study to evaluate the efficacy of nebulised ketamine over normal saline Prior to general anaesthesia in reducing the incidence and severity of post operative sorethroat.

Details of the Patient:

All patients were instructed to fast for at least eight hours prior to surgery. A non-participating anaesthesiologists allocate patients into two groups. The staff nurse opened the envelopes and prepared nebulization solution according to group allocation.

Patients in Group K received ketamine (4.5ml of normal saline and 0.5ml of ketamine 25mg) while patients in Group S received 5ml of normal saline.

Patients received the study drug via nebulization mask connected to an oxygen-driven source (8 L) mounted on the wall for 15 minutes prior to surgery. The registered nurse did not take part in the subsequent evaluation of these patients. Patients were blinded and they were unaware of the nebulisation solution that were administered.

Patient will be shifted to the operating room after 10 minutes of nebulisation and POST score (postoperative sorethroat) will be assessed postoperatively.

Possible risk to the participant/Benefit: the performance of this procedures might help you in reducing the sore throat after surgery.

Compensation to the participant/patient: Nil

Confidentiality: The information obtained in this study is strictly confidential.

Your name or personal identification will not be used in reporting of information in presentations or conferences.

Participants right to withdraw from the study: It's your choice to decide whether to take part in this study or not. If you are not taking part, you don't have to give reason, and it won't affect the care you receive. If you want to take part now, and later you change your mind, you can withdraw from the study at any time.

Complaints regarding the study should be reported to:

The Member Secretary

Institutional Ethics Committee

Saveetha University

162, Poonnamallee High Road

Chennai – 600 077

Phone : 044-66726611

Mobile : 099412 20727

Fax : 044-26800892

Email : dir.res.su@gmail.com

Detailed information and clarification can be obtained from:

Investigator's Name: Jayashree.R

Address: 10/106 Middle street, Meenampatti,

Sivakasi 626189

Mobile: 9677216585

Email : jayashreegowtham09@gmail.com

I, Dr. Jayashree.R (Principal Investigator) have explained clearly to the Participant/patient all the above details. All questions and clarifications by the participant/patient has been fully answered.

Signature of Principal Investigator with date

பங்குபெறுபவர் / நோயாளிதகவல்தாள்

நோயாளியின் தகவல்:

ஆராய்ச்சிதிட்டத்தின் தலைப்பு: நெபுலைஸ் செய்யப்பட்ட கீட்டமைனின் செயல்திறனை மதிப்பிடுவதற்கான ஒரு மருத்துவ ஆய்வு, அறுவைசிகிச்சைக்குப் பின் ஏற்படும் சோர்த்ரோட்டின் நிகழ்வு மற்றும் தீவிரத்தை அளவிடப் போகிறோம்.

ஆய்வு:

ஆய்வுவிளக்கம்: இந்த ஆய்வில் பங்கேற்க உங்கள் அனைவரையும் அழைக்கிறோம், உங்கள் சந்தேகங்களை தெரிந்துகொள்ள இந்த ஆராய்ச்சி பற்றிய முழு தகவல்களும் இந்த தகவல் படிவத்தில் உள்ளது.
அனைத்து நோயாளிகளும் அறுவை சிகிச்சைக்கு எட்டு மணி நேரத்திற்கு முன்பு சாப்பிட அறிவுறுத்தப்பட்டனர். ஒரு மயக்க மருந்து நிபுணர் நோயாளிகளை இரண்டு குழுக்களாகப் பிரிப்பார். பணியாளர் செவிலியர் குழு தகவல் அடங்கிய உறைகளைத் திறந்து குழு ஒதுக்கீட்டின்படி நெபுலைசேஷன் தயாரித்தார்.

K குழுவில் உள்ள நோயாளிகள் கீட்டமைன் நெபுலைசேஷன் பெறுவார்கள், அதே சமயம் S குழுவில் உள்ள நோயாளிகள் சாதாரண சாதாரண உப்பு கரைசல்(normal saline) நெபுலைசேஷன் பெறுவார்கள்.

அறுவைசிகிச்சைக்கு 15 நிமிடங்களுக்கு முன்பு சுவரில் பொருத்தப்பட்ட ஆக்ஸிஜனால் இயக்கப்படும் நெபுலைசேஷன் மாஸ்க் மூலம் நோயாளிகள் ஆய்வு மருந்தைப் பெறுவார்கள். நிர்வகிக்கப்படும் நெபுலைசேஷன் தீர்வு பற்றி நோயாளிகளுக்கு தெரியாது.

10 நிமிட நெபுலைசேஷன் செய்த பிறகு நோயாளி அறுவை சிகிச்சை அறைக்கு மாற்றப்படுவார் மற்றும் அறுவை சிகிச்சைக்குப் பின் POST ஸ்கோர் மதிப்பிடப்படும்.

பங்கேற்பாளருக்கு / நன்மைக்கு சாத்தியமான ஆபத்து:

- இந்த நடைமுறைகளின் செயல்திறன் அறுவை சிகிச்சைக்குப் பிறகு தொண்டை புண்ணைக் குறைக்க உதவும்.

நோயாளிக்குவழங்கப்படும் இழப்பீடு: இழப்பீடு இல்லை

ரகசியக்காப்பு: இந்த ஆய்வில் பெறப்பட்ட தகவல்கள்அனைத்தும் ரகசியமாக இருக்கும்.உங்கள் பெயர் மற்றும் உங்கள் அடையாளங்கள் இல்லாமல் தான் இந்த ஆராய்ச்சியின் முடிவுகளை அறிவிப்போம்

பங்கேற்பாளர்கள் ஆய்வில் இருந்து திரும்பக்கூடிய உரிமை:
இந்த ஆய்வில் பங்கேற்கும்உரிமையும், பங்கேற்க மறுக்கும் உரிமையும், பங்கேற்ற பிறகு திரும்பப் பெறும் உரிமையும்உங்களுக்கு உண்டு. இந்த ஆய்வில் பங்கேற்க உங்கள் அனுமதி தேவைப்படுகிறது.
ஆய்வுதொடர்பான புகார்களை தெரிவிக்கவேண்டிய முகவரி:
உறுப்பினர் செயலர்,
நிறுவனநெறி முறைகள்குழு,
சவீதா பல்கலைக்கழகம்,
162,பூந்தமல்லி ஹைரோடு,
சென்னை - 600077
Phone : 044-66726611
Mobile : 099412 20727
Fax : 044-26800892
Email : dir.res.su@gmail.com

விரிவானதகவல் மற்றும் விளக்கங்களை பெற்றுக்கொள்ள:
டாக்டர். ஜெயஸ்ரீ ஆர்
10/106 நடுத்தர தெரு,
மீனம்பட்டி, சிவகாசி 626189
மொபைல் எண்: 9677216585
மின்னஞ்சல் முகவரி: jayashreegowtham09@gmail.com
நான்,டாக்டர்.ஜெயஸ்ரீ ஆர் மேலேயுள்ள விபரங்கள் அனைத்தையும் தெளிவாக பங்கேற்பாளர் / நோயாளிக்கு விளக்கியுள்ளேன். அனைத்து கேள்விகள் மற்றும் விளக்கங்கள் முழுமையாக பதில் அளிக்கப்பட்டுள்ளது.
கையொப்பம்:
நாள்:

ANNEXURE III

INFORMED CONSENT FORM

I ____ have been invited to take part in the study, conducted by Dr. Jayashree R (PG-Department of Anaesthesiology, Saveetha Medical College and Hospital, Thandalam, Chennai)

Title of the research project: A clinical study to evaluate the efficacy of nebulised ketamine over normal saline Prior to general anesthesia in reducing the incidence and severity of post operative sorethroat.

- I have been explained clearly about the purpose of this study.
- I know about the possible risks of the study.
- I also know about the benefits of this study and the confidentiality of the information during the study.
- I also have the right to refuse or withdraw from the study at any point and the participation in this study is completely voluntary and my consent is required before I can participate in this study.
- I also know whom to contact for complaints or concerns regarding this study, as given in the participation information sheet.

- I acknowledge that I have read the information and same has been explained to me clearly by the principal investigator.
- I agree to give the necessary information's and participate in the study.

Signature of the participant/patient:

Address

Date:

தகவல்படிவம்

நான்____இந்த ஆராய்ச்சிக்கு முழு சம்மதம் தெரிவிக்கிறேன்.

ஆராய்ச்சிதிட்டத்தின்தலைப்பு: நெபுலைஸ் செய்யப்பட்ட கீட்டமைனின் செயல்திறனை மதிப்பிடுவதற்கான ஒரு மருத்துவ ஆய்வு, அறுவைசிகிச்சைக்குப் பின் ஏற்படும் சோர்த்ரோட்டின் நிகழ்வு மற்றும் தீவிரத்தை அளவிடப் போகிறோம்

- இந்த ஆய்வின் நோக்கம் பற்றி எனக்கு தெளிவாக விளக்கப்பட்டுள்ளது
- ஆய்வின் சாத்தியமான அபாயங்கள் பற்றி எனக்குத் தெரியும்.
- இந்த ஆய்வின் நன்மைகள் மற்றும் ஆய்வின் போது தகவல்களின் இரகசியத்தன்மை பற்றியும் எனக்குத் தெரியும்.
- எந்த நேரத்திலும் படிப்பை மறுக்கவோ அல்லது விலகவோ எனக்கு உரிமை உண்டு மேலும் இந்த ஆய்வில் பங்கேற்பது முற்றிலும் தன்னார்வமானது மற்றும் நான் இந்த ஆய்வில் பங்கேற்பதற்கு முன் எனது ஒப்புதல் தேவை.
- பங்கேற்பு தகவல் தாளில் கொடுக்கப்பட்டுள்ளபடி, இந்த ஆய்வு தொடர்பான புகார்கள் அல்லது கவலைகளுக்கு யாரை தொடர்பு கொள்ள வேண்டும் என்பதும் எனக்குத் தெரியும்.
- நான் தகவலைப் படித்தேன் என்பதையும், முதன்மைப் புலனாய்வாளரால் எனக்குத் தெளிவாக விளக்கப்பட்டது என்பதையும் ஒப்புக்கொள்கிறேன்.
- தேவையான தகவல்களை வழங்கவும், ஆய்வில் பங்கேற்கவும் ஒப்புக்கொள்கிறேன்.

கையொப்பம் :

முகவரி :

தேதி:

ANNEXURE – IV

LIST OF FIGURES

FIG NO:	FIGURE NAME
1	PARTS OF ENDOTRACHEAL TUBE
2	PICTORIAL REPRESENTATION OF TWO TYPES OF CUFFS
3	HIGH VOLUME LOW PRESSURE AND LOW VOLUME HIGH PRESSURE CUFF AFTER INFLATION
4	EFFECT OF CUFF ON TRACHEAL MUCOSA
5	CUFF PRESSURE MANOMETER
6	EFFECT OF HARD CUFF AND SOFT CUFF ON TRACHEAL MUCOSA
7	MECHANISM OF ACTION OF KETAMINE
8	ACTIONS ON KETAMINE ON VARIOUS CHANNELS
9	PRESSURIZED NEBULISER FROM 1858
10	PARTS OF NEBULISER
11	CONSOLIDATED STANDARD OF REPORTING TRIALS (CONSORT)
12	AGE DISTRIBUTION OF STUDY PARTICIPANTS
13	GENDER DISTRIBUTION OF STUDY PARTICIPANTS
14	BODY MASS INDEX DISTRIBUTION OF STUDY PARTICIPANTS
15	DURATION OF SURGERY DISTRIBUTION OF STUDY

	PARTICIPANTS
16	CHANGES IN HEART RATE AMONG THE STUDY PARTICIPANTS
17	CHANGES IN SYSTOLIC BLOOD PRESSURE AMONG THE STUDY PARTICIPANTS
18	CHANGES IN DIASTOLIC BLOOD PRESSURE AMONG THE STUDY PARTICIPANTS
19	CHANGES IN MEAN ARTERIAL PRESSURE AMONG THE STUDY PARTICIPANTS
20	AFTER EXTUBATION IN OT POST SCORE DISTRIBUTION IN STUDY PARTICIPANTS
21	RECOVERY ROOM POST SCORE "0" HOUR distribution in study participants
22	POST SCORE "2nd" HOUR distribution in study participants
23	POST SCORE "4th" HOUR distribution in study participants
24	POST SCORE "8th" HOUR distribution in study participants
25	POST SCORE "12th" HOUR distribution in study participants
26	POST SCORE "24th" HOUR distribution in study participants

ANNEXURE – V

LIST OF TABLES

TABLE NUMBER	TABLE NAME
1	AGE DISTRIBUTION OF THE STUDY PARTICIPANTS
2	GENDER DISTRIBUTION OF THE STUDY PARTICIPANTS
3	BODY MASS INDEX AND DURATION OF SURGERY OF STUDY PARTICIPANTS
4	DISTRIBUTION OF HEART RATE AMONG THE STUDY PARTICIPANTS
5	DISTRIBUTION OF SYSTOLIC BLOOD PRESSURE AMONG THE STUDY PARTICIPANTS
6	DISTRIBUTION OF DIASTOLIC BLOOD PRESSURE AMONG THE STUDY PARTICIPANTS
7	DISTRIBUTION OF MEAN ARTERIAL PRESSURE AMONG THE STUDY PARTICIPANTS
8	DISTRIBUTION OF OXYGEN SATURATION AMONG THE STUDY PARTICIPANTS
9	DISTRIBUTION OF AFTER EXTUBATION IN OT POST SCORE AMONG THE STUDY PARTICIPANTS
10	DISTRIBUTION OF RECOVERY ROOM “0” HOUR POST SCORE AMONG THE STUDY PARTICIPANTS.

11	DISTRIBUTION OF POST SCORE "2nd" HOUR AMONG THE STUDY PARTICIPANTS
12	DISTRIBUTION OF POST SCORE "4th" HOUR AMONG THE STUDY PARTICIPANTS
13	DISTRIBUTION OF POST SCORE "8th" HOUR AMONG THE STUDY PARTICIPANTS
14	DISTRIBUTION OF POST SCORE "12th" HOUR AMONG THE STUDY PARTICIPANTS
15	DISTRIBUTION OF POST SCORE "24th" HOUR AMONG THE STUDY PARTICIPANTS.

ANNEXURE – VI

LIST OF ABBREVIATIONS

ABBREVIATION	EXPANSION
POST	POSTOPERATIVE SORETHROAT
GA	GENERAL ANAESTHESIA
ETT	ENDOTRACHEAL TUBE
NSAID	NON-STEROIDAL ANTI-INFLAMMATORY DRUGS
PVC	POLY VINYL CHLORIDE
BMI	BODY MASS INDEX
ECG	ELECTROCARDIOGRAM
ASA	AMERICAN SOCEITY OF ANAESTHESIOLOGIST
PACU	POST ANAESTHESIA CARE UNIT
PREMED	PREMEDICATION
NMDA	N-METHYL-D-ASPARTATE RECEPTOR ANTAGONIST

ANNEXURE – VII : MASTER CHART

S.No	AGE	Sex	GROUP	HEIGHT	WEIGHT	BMI	ASA	PROCEDURE	TYPE OF ANESTHESIA	Duration of surgery (min)	PREMED HR	SPO2	SBP	DBP	MAP	PRE IND HR	SPO2	SBP	DBP	MAP	INDUCTION HR	SPO2	SBP	DBP	MAP	15 MIN HR	SPO2	SBP	DBP	MAP	30 MIN HR	SPO2	SBP	DBP	MAP	45 MIN HR	SPO2	SBP	DBP	MAP	AFTER EXTUBATION POST SCORE(OT)	RECOVERY ROOM O HOUR	2ND HOUR	4TH HOUR	8TH HOUR	12TH HOUR	24TH HOUR
1	34	Male	S	157	81	32.8	1	ORIF CLAVILCE LEFT	General anesthesia	82	88	98	116	83	94	93	99	106	82	90	78	100	131	74	93	70	100	96	69	78	87	100	127	74	91.6	82	100	130	90	103.3	2	2	2	2	1	1	0
2	21	Male	K	158	58	23.2	1	ORIF CLAVILCE LEFT	General anesthesia	78	98	98	128	75	92.6	92	100	117	76	89.6	103	100	126	84	98	95	100	98	83	88	62	100	116	82	93.3	73	100	128	67	87.3	0	0	0	0	0	0	0
3	11	Male	S	161	57	21.9	2	LEFT PCNL	General anesthesia	114	99	99	119	79	92.3	68	100	120	70	86.6	113	100	131	98	109	77	100	117	85	95.6	60	100	116	80	92	73	100	122	87	98.6	1	1	1	1	1	0	0
4	65	Male	S	166	60	21.77384	2	ERCP	General anesthesia	161	89	100	113	73	86.3	87	99	108	62	77.3	94	100	131	78	95.6	65	100	117	83	94.3	88	100	121	65	83.6	81	100	128	72	90.6	2	2	2	2	2	1	0
5	57	Male	S	162	75	28.5	2	laprotomy	General anesthesia	139	77	100	116	80	92	106	99	108	75	86	91	100	135	85	101.6	64	100	110	77	88	61	100	128	67	87.3	73	100	120	97	104.6	0	0	0	0	0	0	0
6	17	Female	K	162	57	21.7	2	LSCS	General anesthesia	74	90	98	115	72	86.3	110	99	120	84	96	113	100	120	71	87.3	97	100	91	84	86.3	59	100	114	69	84	86	100	120	91	100.6	0	0	0	0	0	0	0
7	33	Male	S	165	81	29.7	2	Microlaryngeal excision of lesion	General anesthesia	132	65	98	114	70	84.6	87	99	114	66	82	119	100	135	84	101	63	100	119	71	87	72	100	114	77	89.3	85	100	130	88	102	0	0	0	0	0	0	0
8	25	Male	K	164	71	26.3	1	lap appendicectomy	General anesthesia	118	61	99	117	67	83.6	82	100	113	73	86.3	85	100	130	77	94.6	91	100	94	66	75.3	60	100	125	81	95.6	75	100	122	74	90	1	1	0	0	0	0	0
9	36	Male	S	165	59	21.6	1	lap appendicectomy	General anesthesia	149	94	100	114	82	92.6	68	99	110	82	91.3	93	100	122	98	106	85	100	91	75	80.3	77	100	127	70	89	72	100	127	97	107	1	1	1	1	1	0	0
10	68	Female	S	165	87	31.9	1	lap TEPP	General anesthesia	84	78	98	126	69	88	106	100	128	74	92	95	100	134	97	109.3	65	100	99	88	91.6	88	100	118	67	84	85	100	122	84	96.6	0	0	0	0	0	0	0
11	60	Male	S	158	70	28	2	laparotomy	General anesthesia	145	92	100	125	79	94.3	89	99	127	80	95.6	77	100	134	85	101.3	61	100	122	78	92.6	65	100	118	68	84.6	86	100	123	95	104.3	0	0	0	0	0	0	0
12	74	Male	K	168	83	29.4	2	laparotomy	General anesthesia	165	86	98	112	77	88.6	95	99	125	76	92.3	95	100	132	89	103.3	85	100	96	75	82	89	100	115	85	95	78	100	130	100	110	1	1	1	1	1	0	0
13	29	Female	S	158	83	33.2	2	laparotomy	General anesthesia	99	83	100	125	84	97.6	65	100	126	64	84.6	99	100	121	98	105.6	76	100	107	69	81.6	73	100	127	77	93.6	81	100	126	74	91.3	1	1	1	1	1	1	0
14	36	Male	S	162	86	32.7	3	bilateral lap adrenalectomy	General anesthesia	179	63	100	113	67	82.3	64	99	127	86	99.6	79	100	121	73	89	90	100	91	68	75.6	87	100	123	88	99.6	77	100	130	92	104.6	2	2	2	1	1	1	1
15	49	Male	S	167	67	24.02381	3	partial penectomy	General anesthesia	136	63	98	112	82	92	67	99	127	69	88.3	87	100	135	82	99.6	80	100	119	83	95	87	100	118	65	82.6	86	100	123	80	94.3	0	0	0	0	0	0	0
16	52	Female	S	167	66	23.66524	2	Right RIRS	General anesthesia	111	100	99	118	88	98	76	100	128	78	94.6	108	100	130	80	96.6	71	100	118	88	98	86	100	118	76	90	88	100	128	70	89.3	1	1	1	1	1	1	0
17	74	Male	K	157	70	28.3	2	cystoscopy and proceed	General anesthesia	179	65	98	130	80	96.6	101	100	126	61	82.6	84	100	135	71	92.3	72	100	104	83	90	90	100	115	78	90.3	90	100	121	91	101	0	0	0	0	0	0	0
18	20	Male	S	166	87	31.5	2	perineal urethroplasty	General anesthesia	169	84	98	115	90	98.3	63	100	105	85	91.6	119	100	135	85	101.6	77	100	111	73	85.6	69	100	126	84	98	72	100	125	95	105	0	0	0	0	0	0	0
19	50	Female	S	162	68	25.9	2	left RIRS	General anesthesia	119	88	98	130	66	87.3	75	99	126	66	86	83	100	128	89	102	62	100	98	75	82.6	82	100	127	67	87	85	100	129	89	102.3	0	0	0	0	0	0	0
20	52	Male	K	166	71	25.7	2	left PCNL	General anesthesia	64	88	99	112	76	88	79	99	103	78	86.3	119	100	130	94	106	89	100	115	75	88.3	73	100	118	71	86.6	86	100	126	83	97.3	1	1	0	0	0	0	0
21	78	Male	S	159	56	22.1	3	nephrectomy left side	General anesthesia	67	63	99	114	89	97.3	72	100	108	79	88.6	91	100	120	77	91.3	86	100	90	67	74.6	86	100	127	86	99.6	75	100	122	76	91.3	0	0	0	0	0	0	0
22	19	Female	S	158	80	32	3	posterior stabilisation D8-D11	General anesthesia	75	65	99	122	65	84	64	99	107	86	93	83	100	126	95	105.3	97	100	93	79	83.6	79	100	118	80	92.6	82	100	124	72	89.3	1	1	1	1	1	0	0
23	46	Male	S	156	66	27.1	3	L4-L5 laminectomy	General anesthesia	159	88	98	128	80	96	100	99	118	67	84	89	100	120	86	97.3	96	100	116	69	84.6	84	100	114	77	89.3	92	100	123	97	105.6	0	0	0	0	0	0	0
24	50	Male	S	165	78	28.6	3	L4-L5 laminectomy	General anesthesia	113	76	100	114	79	90.6	74	99	119	81	93.6	83	100	133	72	92.3	87	100	122	77	92	67	100	124	65	84.6	82	100	130	98	108.6	0	0	0	0	0	0	0
25	46	Male	K	168	77	27.2	3	laminectomy and disectomy	General anesthesia	117	84	100	117	85	95.6	105	99	120	81	94	118	100	130	84	99.3	86	100	122	77	92	83	100	118	75	89.3	79	100	128	86	100	1	1	0	0	0	0	0
26	49	Male	S	164	64	23.7	3	D8-D11 posterior stabilisation	General anesthesia	96	78	99	130	72	91.3	109	100	103	74	83.6	115	100	123	84	97	71	100	98	69	78.6	84	100	125	70	88.3	82	100	123	92	102.3	2	2	2	2	2	2	1
27	58	Male	K	160	69	26.9	3	C3-C4 ACDF	General anesthesia	113	85	100	115	77	89.6	62	99	101	71	81	83	100	126	94	104.6	76	100	105	84	91	76	100	122	66	84.6	77	100	121	67	85	0	0	0	0	0	0	0
28	61	Male	S	166	57	20.6	3	C2-C6 ACDF	General anesthesia	68	97	100	129	83	98.3	77	99	115	84	94.3	76	100	124	82	96	80	100	100	88	92	71	100	126	75	92	85	100	124	81	95.3	1	1	1	1	0	0	0
29	42	Male	S	168	71	25.1	3	C6-C7 ACDF	General anesthesia	105	74	99	129	69	89	61	100	116	86	96	96	100	135	99	111	92	100	107	85	92.3	66	100	119	73	88.3	77	100	122	92	102	0	0	0	0	0	0	0
30	32	Female	K	160	56	21.8	1	excision of fibroadenoma	General anesthesia	113	88	100	112	74	86.6	103	99	109	76	87	94	100	133	96	108.3	62	100	112	87	95.3	58	100	112	79	90	78	100	127	100	109	0	0	0	0	0	0	0
31	43	Male	S	157	85	34.4	1	lap cholecystectomy	General anesthesia	166	93	100	118	89	98.6	88	99	121	76	91	79	100	126	71	89.3	93	100	93	89	90.3	58	100	117	80	92.3	76	100	130	79	96	2	2	2	2	2	1	1
32	26	Female	S	163	81	30.4	2	tendon release	General anesthesia	151	94	98	128	77	94	68	100	116	78	90.6	93	100	122	77	92	61	100	123	84	97	89	100	128	87	100.6	73	100	122	67	85.3	3	3	3	3	2	2	1
33	25	Male	S	168	71	25.1	1	lap appendicectomy	General anesthesia	156	100	99	120	75	90	62	100	106	70	82	72	100	135	93	107	96	100	102	74	83.3	78	100	127	74	91.6	77	100	121	95	103.6	0	0	0	0	0	0	0
34	26	Male	S	163	84	31.6	2	left leg wound debridement	General anesthesia	165	89	99	128	84	98.6	76	99	108	78	88	89	100	126	87	100	61	100	122	66	84.6	89	100	119	77	91	84	100	121	84	96.3	1	1	1	1	1	1	1
35	53	Male	K	164	78	29	1	left shoulder rotation flap	General anesthesia	120	74	98	126	68	87.3	65	99	124	83	96.6	103	100	127	74	91.6	72	100	121	82	95	77	100	122	81	94.6	87	100	126	91	102.6	2	2	2	2	1	1	0
36	52	Male	S	158	62	24.8	1	lap cholecystectomy	General anesthesia	167	79	98	112	69	83.3	80	99	102	62	75.3	116	100	135	74	94.3	71	100	97	86	89.6	90	100	118	75	89.3	85	100	127	68	87.6	0	0	0	0	0	0	0
37	42	Male	K	158	58	23.2	1	lap meshplasty	General anesthesia	88	96	100	124	81	95.3	91	100	101	87	91.6	83	100	131	77	95	73	100	122	80	94	76	100	118	87	97.3	80	100	130	70	90	0	0	0	0	0	0	0

38	20	Male	S	168	82	29	1	d lap and proceed	General anesthesia	130	97	99	122	79	93.3	110	99	114	87	96	114	100	124	71	88.6	93	100	100	70	80	85	100	116	78	90.6	84	100	122	98	106	2	2	2	2	1	1	0
39	55	Male	K	164	67	24.9	2	carbuncle back excision	General anesthesia	117	95	98	118	79	92	97	99	113	82	92.3	74	100	122	77	92	61	100	111	65	80.3	59	100	118	70	86	80	100	121	68	85.6	1	1	0	0	0	0	0
40	61	Female	S	165	87	31.9	3	free flap fibula	General anesthesia	71	89	99	116	78	90.6	61	100	124	69	87.3	78	100	120	95	103.3	89	100	96	88	90.6	85	100	118	67	84	82	100	124	98	106.6	1	1	1	1	1	1	1
41	37	Female	S	167	74	26.5	2	tangential excision of burns	General anesthesia	146	63	99	126	90	102	91	99	128	73	91.3	116	100	125	86	99	94	100	96	68	77.3	74	100	128	73	91.3	83	100	122	98	106	0	0	0	0	0	0	0
42	55	Male	S	164	80	29.7	2	d lap and proceed	General anesthesia	160	63	99	120	71	87.3	74	100	113	80	91	105	100	129	75	93	81	100	111	74	86.3	75	100	120	67	84.6	77	100	120	69	86	0	0	0	0	0	0	0
43	57	Female	S	157	85	34.4	2	carbuncle back excision	General anesthesia	146	74	100	123	65	84.3	91	99	103	69	80.3	73	100	125	83	97	72	100	94	76	82	63	100	123	66	85	73	100	120	85	96.6	1	1	1	0	0	0	0
44	23	Male	S	164	66	24.5	2	ORIF radius	General anesthesia	69	83	100	118	73	88	90	99	105	79	87.6	109	100	124	84	97.3	85	100	104	86	92	79	100	116	86	96	83	100	130	82	98	0	0	0	0	0	0	0
45	26	Female	K	156	72	29.5	2	incision and drainage breast right side	General anesthesia	139	98	100	116	78	90.6	94	99	106	74	84.6	104	100	131	81	97.6	69	100	99	83	88.3	65	100	118	65	82.6	81	100	125	81	95.6	0	0	0	0	0	0	0
46	64	Male	S	158	62	24.8	2	incision and drainage	General anesthesia	90	95	99	125	66	85.6	75	99	124	79	94	117	100	134	93	106.6	84	100	95	82	86.3	61	100	121	87	98.3	77	100	124	71	88.6	0	0	0	0	0	0	0
47	52	Male	K	162	81	30.8	2	lap cholecystectomy	General anesthesia	76	82	98	125	82	96.3	108	99	120	81	94	100	100	122	89	100	75	100	117	85	95.6	76	100	118	82	94	80	100	128	95	106	1	1	1	1	1	1	1
48	18	Female	S	158	68	27.2	2	lap meshplasty	General anesthesia	117	78	100	125	78	93.6	61	100	126	77	93.3	74	100	125	83	97	76	100	102	85	90.6	81	100	126	84	98	88	100	128	91	103.3	0	0	0	0	0	0	0
49	30	Female	S	167	80	28.6	1	lap appendicectomy	General anesthesia	79	70	100	130	86	100.6	68	99	116	70	85.3	77	100	131	81	97.6	63	100	123	68	86.3	81	100	117	69	85	88	100	127	69	88.3	2	2	2	2	1	0	0
50	20	Female	S	161	76	29.3	2	laparotomy	General anesthesia	109	64	99	127	76	93	90	100	114	72	86	107	100	133	99	110.3	87	100	103	76	85	65	100	115	67	83	72	100	127	92	103.6	0	0	0	0	0	0	0
51	20	Male	S	156	72	29.5	2	laparotomy	General anesthesia	60	70	99	120	84	96	103	100	114	83	93.3	100	100	135	92	106.3	68	100	104	89	94	60	100	113	78	89.6	83	100	123	91	101.6	2	2	2	2	2	2	1
52	75	Male	S	157	64	25.9	2	TURP	General anesthesia	155	97	100	128	67	87.3	82	100	109	72	84.3	106	100	132	70	90.6	67	100	92	70	77.3	67	100	118	68	84.6	75	100	121	97	105	2	2	2	2	2	1	0
53	43	Male	K	157	87	35.2	3	DJ stent removal	General anesthesia	85	83	98	121	84	96.3	75	99	128	85	99.3	90	100	134	70	91.3	76	100	113	89	97	58	100	123	68	86.3	89	100	128	76	93.3	0	0	0	0	0	0	0
54	18	Male	S	156	77	31.6	1	flexibile urethroscopy	General anesthesia	145	97	98	118	85	96	66	99	104	68	80	83	100	131	86	101	76	100	109	86	93.6	69	100	119	66	83.6	85	100	120	89	99.3	3	3	3	3	3	2	2
55	18	Female	S	165	62	22.7	2	right PCNL	General anesthesia	88	71	99	130	75	93.3	73	100	124	70	88	73	100	125	86	99	81	100	93	77	82.3	68	100	122	70	87.3	78	100	126	97	106.6	2	2	2	2	2	1	1
56	76	Female	K	156	60	24.65483	2	hysterectomy	General anesthesia	68	69	98	128	81	96.6	96	100	110	63	78.6	114	100	132	97	108.6	67	100	104	82	89.3	64	100	123	74	90.3	91	100	121	71	87.6	1	1	1	1	0	0	0
57	68	Female	S	158	59	23.63403	2	lap cholecystectomy	General anesthesia	163	84	99	117	66	83	70	100	107	67	80.3	71	100	128	77	94	71	100	114	70	84.6	58	100	127	80	95.6	87	100	128	78	94.6	0	0	0	0	0	0	0
58	44	Female	S	158	69	27.6398	1	excision and biopsy	General anesthesia	140	85	100	124	69	87.3	77	99	128	72	90.6	115	100	121	74	89.6	67	100	93	74	80.3	69	100	121	83	95.6	86	100	120	68	85.3	0	0	0	0	0	0	0
59	40	Female	S	157	64	25.96454	2	robotic and lap hysterectomy	General anesthesia	163	85	100	124	69	87.3	76	99	108	61	76.6	96	100	128	80	96	90	100	115	74	87.6	83	100	124	75	91.3	78	100	129	78	95	0	0	0	0	0	0	0
60	38	Female	S	166	69	25.03992	1	breast conservative surgery	General anesthesia	140	67	98	124	81	95.3	106	99	119	80	93	85	100	122	92	102	78	100	116	78	90.6	83	100	122	79	93.3	80	100	125	66	85.6	0	0	0	0	0	0	0
61	57	Female	S	165	67	24.60973	2	wound debridement	General anesthesia	85	81	99	124	70	88	105	100	115	85	95	112	100	120	81	94	93	100	123	78	93	85	100	114	85	94.6	92	100	123	84	97	2	2	2	1	1	0	0
62	56	Male	S	166	55	19.95936	3	ileostomy	General anesthesia	138	94	100	115	79	91	79	99	118	82	94	99	100	132	91	104.6	78	100	109	87	94.3	63	100	121	80	93.6	83	100	126	82	96.6	0	0	0	0	0	0	0
63	20	Female	S	156	81	33.28402	2	lap cholecystectomy	General anesthesia	173	95	100	114	77	89.3	79	100	127	60	82.3	84	100	131	92	105	94	100	118	82	94	69	100	121	70	87	75	100	121	68	85.6	1	1	1	0	0	0	0
64	60	Female	K	157	63	25.55885	2	lap cholecystectomy	General anesthesia	123	68	100	114	90	98	110	99	121	70	87	95	100	132	78	96	67	100	122	75	90.6	77	100	112	73	86	71	100	130	92	104.6	0	0	0	0	0	0	0
65	63	Male	S	162	87	33.15043	2	wound debridement	General anesthesia	135	96	100	116	74	88	70	99	113	79	90.3	87	100	132	96	108	91	100	114	78	90	76	100	125	87	99.6	87	100	126	90	102	2	2	2	1	0	0	0
66	76	Male	S	164	74	27.51338	3	WLE and transposition flap	General anesthesia	130	69	99	124	73	90	104	99	110	62	78	101	100	121	96	104.3	81	100	117	75	89	65	100	116	78	90.6	80	100	125	74	91	0	0	0	0	0	0	0
67	37	Female	S	162	83	31.62628	3	robotic assisted ovarian cystectomy	General anesthesia	109	96	100	114	72	86	85	100	103	73	83	108	100	130	97	108	69	100	104	72	82.6	63	100	127	85	99	91	100	122	76	91.3	0	0	0	0	0	0	0
68	54	Female	K	159	77	30.45766	3	robotic assisted cholecystectomy	General anesthesia	155	88	99	130	66	87.3	102	99	116	67	83.3	100	100	120	86	97.3	89	100	112	78	89.3	88	100	126	87	100	86	100	130	85	100	0	0	0	0	0	0	0
69	41	Female	K	158	60	24.03461	2	lap cholecystectomy	General anesthesia	138	100	100	123	72	89	85	100	109	76	87	74	100	126	82	96.6	71	100	107	72	83.6	62	100	117	77	90.3	87	100	122	71	88	1	1	1	1	1	0	0
70	32	Female	K	168	85	30.11621	3	left breast MRM	General anesthesia	105	82	98	112	66	81.3	62	100	120	63	82	78	100	126	82	96.6	77	100	102	83	89.3	85	100	127	75	92.3	92	100	130	78	95.3	0	0	0	0	0	0	0
71	70	Female	K	157	58	23.53037	3	WLE of shoulder squamous ca	General anesthesia	121	69	100	112	74	86.6	106	100	100	63	75.3	92	100	135	81	99	62	100	123	75	91	86	100	118	71	86.6	85	100	121	74	89.6	0	0	0	0	0	0	0
72	72	Male	K	159	72	28.47989	3	gastrectomy for ca stomach	General anesthesia	88	74	98	118	74	88.6	77	100	116	74	88	110	100	120	72	88	77	100	104	70	81.3	88	100	122	82	95.3	81	100	128	92	104	0	0	0	0	0	0	0
73	50	Female	K	161	71	27.39092	3	vaginal hysterectomy	General anesthesia	144	91	100	123	66	85	97	100	127	60	82.3	80	100	124	73	90	87	100	119	87	97.6	82	100	112	72	85.3	89	100	127	96	106.3	0	0	0	0	0	0	0
74	52	Female	K	158	77	30.84442	3	TAH with BSO	General anesthesia	125	92	98	120	86	97.3	74	100	124	86	98.6	95	100	133	87	102.3	72	100	105	82	89.6	60	100	124	69	87.3	79	100	123	83	96.3	0	0	0	0	0	0	0
75	49	Male	K	162	71	27.0538	2	lap cholecystectomy	General anesthesia	121	61	98	114	88	96.6	66	99	123	80	94.3	85	100	132	88	102.6	78	100	122	65	84	72	100	118	78	91.3	76	100	129	95	106.3	0	0	0	0	0	0	0
76	45	Female	K	164	70	26.02617	3	TLIF	General anesthesia	63	72	98	121	75	90.3	84	100	110	87	94.6	115	100	122	89	100	94	100	123	87	99	83	100	120	78	92	86	100	130	86	100.6	2	2	2	1	1	1	0
77	57	Female	K	159	67	26.50212	3	L3-L5 laminectomy	General anesthesia	175	94	99	118	86	96.6	61	99	123	74	90.3	93	100	133	85	101	74	100	92	76	81.3	89	100	124	67	86	81	100	121	81	94.3	0	0	0	0	0	0	0
78	51	Male	K	161	60	23.14726	3	L4-L5 laminectomy	General anesthesia	93	80	98	112	75	87.3	84	99	101	85	90.3	95	100	131	96	107.6	98	100	98	73	81.3	73	100	118	80	92.6	72	100	122	85	97.3	0	0	0	0	0	0	0
79	67	Male	K	163	61	22.95909	3	TLIF	General anesthesia	113	73	99	113	88	96.3	89	100	106	60	75.3	77	100	130	82	98	75	100	92	83	86	85	100	120	86	97.3	92	100	127	99	108.3	0	0	0	0	0	0	0
80	41	Male	K	163	86	32.36855	2	left diagnostic arthroscopy shoulder	General anesthesia	71	83	98	127	78	94.3	94	100	109	74	85.6	86	100	125	72	89.6	94	100	105	78	87	65	100	113	85	94.3	91	100	123	89	100.3	0	0	0	0	0	0	0
81	40	Male	K	163	71	26.72287	2	left diagnostic arthroscopy shoulder	General anesthesia	98	65	100	130	68	88.6	73	99	123	66	85	120	100	134	84	100.6	88	100	98	77	84	86	100	114	66	82	80	100	129	66	87	0	0	0	0	0	0	0
82	40	Female	K	164	86	31.97501	3	TLIF	General anesthesia	164	88	99	114	84	94	76	100	120	66	84	99	100	127	71	89.6	76	100	96	74	81.3	72	100	127	84	98.3	77	100	130	97	108	2	2	1	1	0	0	0
83	24	Male	K	159	67	26.50212	3	Mini discectomy	General anesthesia	131	91	100	120	71	87.3	94	99	127	80	95.6	83	100	124	72	89.3	99	100	108	65	79.3	78	100	125	65	85	76	100	122	69	86.6	0	0	0	0	0	0	0
84	62	Female	K	166	84	30.48338	3	right hemicolectomy	General anesthesia	69	98	98	118	84	95.3	107	99	117	71	86.3	113	100	134	84	100.6	79	100	95	81	85.6	81	100	124	79	94	76	100	124	76	92	0	0	0	0	0	0	0

85	25	Female	K	163	74	27.85201	2	lap cholecystectomy	General anesthesia	136	91	98	116	67	83.3	104	100	101	61	74.3	114	100	122	75	90.6	76	100	94	65	74.6	63	100	121	79	93	73	100	127	89	101.6	0	0	0	0	0	0	0
86	47	Male	K	167	86	30.83653	3	incision and drainage	General anesthesia	128	83	98	115	81	92.3	73	99	110	82	91.3	112	100	131	88	102.3	72	100	107	88	94.3	58	100	116	83	94	70	100	127	100	109	0	0	0	0	0	0	0
87	50	Female	K	164	60	22.30815	1	lap cholecystectomy	General anesthesia	86	88	100	117	87	97	90	99	104	87	92.6	84	100	135	99	111	60	100	110	72	84.6	58	100	118	78	91.3	89	100	129	67	87.6	0	0	0	0	0	0	0
88	54	Female	K	156	74	30.40763	1	lap cholecystectomy	General anesthesia	171	87	100	123	68	86.3	96	100	124	80	94.6	82	100	134	76	95.3	61	100	104	72	82.6	77	100	120	70	86.6	85	100	121	96	104.3	0	0	0	0	0	0	0
89	25	Female	K	166	64	23.22543	1	lap cholecystectomy	General anesthesia	176	67	100	119	85	96.3	68	99	110	74	86	83	100	120	73	88.6	95	100	92	65	74	82	100	119	72	87.6	91	100	125	92	103	0	0	0	0	0	0	0
90	41	Male	K	168	68	24.09297	1	d lap and proceed	General anesthesia	70	98	100	118	85	96	79	100	122	63	82.6	84	100	130	82	98	69	100	110	85	93.3	78	100	127	72	90.3	77	100	125	66	85.6	0	0	0	0	0	0	0
91	56	Female	K	157	60	24.34176	3	open LAR	General anesthesia	92	100	99	122	73	89.3	94	99	116	74	88	102	100	127	96	106.3	91	100	105	87	93	68	100	119	79	92.3	90	100	123	66	85	0	0	0	0	0	0	0
92	71	Male	K	162	79	30.10212	1	lap cholecystectomy	General anesthesia	156	79	100	129	84	99	61	100	125	87	99.6	71	100	131	79	96.3	83	100	106	88	94	75	100	127	72	90.3	80	100	123	94	103.6	0	0	0	0	0	0	0
93	63	Male	K	161	85	32.79194	2	lap TEPP	General anesthesia	150	95	98	119	77	91	69	100	105	68	80.3	119	100	122	92	102	96	100	104	70	81.3	58	100	112	70	84	86	100	120	96	104	2	2	2	2	1	0	0
94	67	Female	K	159	56	22.15102	2	lap meshplasty	General anesthesia	175	66	98	123	69	87	103	100	120	61	80.6	70	100	128	73	91.3	99	100	119	65	83	66	100	113	71	85	71	100	127	88	101	0	0	0	0	0	0	0
95	29	Female	K	168	85	30.11621	1	right breast fibroadenoma	General anesthesia	156	89	99	126	79	94.6	103	100	123	75	91	116	100	134	93	106.6	72	100	106	76	86	80	100	117	68	84.3	78	100	124	99	107.3	0	0	0	0	0	0	0
96	31	Female	K	166	78	28.306	1	lap appendicectomy	General anesthesia	172	85	98	118	70	86	83	100	121	73	89	78	100	134	73	93.3	80	100	123	75	91	73	100	118	77	90.6	90	100	130	96	107.3	0	0	0	0	0	0	0
97	18	Male	K	158	73	29.24211	1	lap appendicectomy	General anesthesia	79	86	100	128	71	90	90	100	103	70	81	71	100	123	95	104.3	88	100	90	73	78.6	84	100	122	82	95.3	75	100	129	94	105.6	0	0	0	0	0	0	0
98	47	Female	K	165	72	26.44628	1	lap appendicectomy	General anesthesia	124	81	98	112	66	81.3	68	99	113	82	92.3	120	100	133	83	99.6	61	100	105	78	87	81	100	126	87	100	76	100	122	69	86.6	0	0	0	0	0	0	0
99	20	Female	K	165	61	22.40588	1	excision of fibroadenoma	General anesthesia	167	60	100	118	69	85.3	106	100	124	64	84	86	100	128	70	89.3	76	100	99	89	92.3	73	100	116	66	82.6	78	100	127	88	101	0	0	0	0	0	0	0
100	56	Male	K	167	77	27.60945	2	carbuncle back excision	General anesthesia	127	95	98	128	86	100	86	99	116	63	80.6	76	100	125	83	97	73	100	117	70	85.6	68	100	116	74	88	92	100	120	100	106.6	0	0	0	0	0	0	0

www.ingramcontent.com/pod-product-compliance
Lightning Source LLC
LaVergne TN
LVHW082356200926
848481LV00012B/3584